Every Page is Page One

Topic-based Writing for Technical Communication and the Web

Mark Baker

Every Page is Page One

Topic-based Writing for Technical Communication and the Web

Credits

Cover Image:	Wall Assemblage – Copyright © 2006 Eric Lin on flickr (CC BY-SA 2.0)
Cover Background:	Another Brick in the Wall – Copyright © 2009 Coun2rparts on flickr (CC BY-SA 2.0)
Foreword:	Scott Abel, The Content Wrangler

Disclaimer

Trademarks

XML Press
Laguna Hills, California 92637
http://xmlpress.net

First Edition
ISBN: 978-1-937434-28-1 (print)
ISBN: 978-1-937434-29-8 (ebook)

Table of Contents

Foreword vii
Preface: In the Context of the Web ix
  Audience x
  Discussion xi
  Acknowledgments xi
1. Introduction 1
  Every page is page one 3
  Why do we still write books? 4
  About the book 6

I. Content in the Context of the Web 7

  2. Include it all. Filter it afterward. 9
    Just Google it 10
    The long tail 11
    Authority and experience 17
    Aggregation and curation 20
    Filter it afterward 21
  3. The Distributed Nature of Content on the Web 25
    How we use the Web 25
    Dynamic semantic clustering 26
  4. Information Architecture Top Down 31
    Book navigation 31
    The trouble with TOCs 32
    Curriculum versus classification 34
    The limits of hierarchies 38
    The cultural bias toward hierarchies 39
    The rise of the Frankenbooks 41
    Faceted navigation 43
    The limits of classification 45
    Where top-down works 47
  5. Information Architecture Bottom Up 51
    A web of subject affinities 53
    Irregular subject affinities 58
    Subject affinities are not citations 59

Topics as hubs 60
The flattening problem 60
Broader, deeper, more dynamic 64
Should we abandon top-down navigation? 65
The role of lists 66

II. Characteristics of Every Page is Page One Topics 69

6. What is a Topic? 71
Building-block topics 71
Presentational topics 73
Every Page is Page One topics 73
Economics and the evolution of topics 75
DITA and Information Mapping 75
Topics and the Web 76
Every page is still page one even if the reader reads several 77
Characteristics of EPPO topics 78

7. EPPO Topics are Self-contained 79
Self-contained, not all alone 82
The information scent of self-contained topics 83

8. EPPO Topics have a Specific and Limited Purpose 85
The scope of a topic 85
Task-based writing 86
Derived purpose 88
Defining the purpose of a topic 89
Topic purpose vs. user purpose 90
Purpose and topic size 92
Decision support and the reader's purpose 92
Purpose and findability 95

9. EPPO Topics Conform to a Type 97
The evolution of topic types 100
Discovering and defining topic types 102
Concept, task, and reference reconsidered 105

10. EPPO Topics Establish their Context 117
Establishing context 118
Context and the imprecision of search 121

11. EPPO Topics Assume the Reader is Qualified 123
Reader dependencies vs. subject dependencies 125
Determining the qualified reader 127
Choosing the level of understanding 127

Avoid arbitrary labels 129
Qualification and findability 129
12. EPPO Topics Stay on One Level 131
Books change levels at the author's fiat 132
Keeping topics on one level 135
13. EPPO Topics Link Richly 139
Links and the democratization of knowledge 141
Linking and findability 142
III. Writing Every Page is Page One Topics 145
14. Writing Every Page is Page One Topics 147
Textbooks vs. user assistance 147
Writing topics 150
The question of style 159
Concerning reference information 161
Concerning tutorials 162
Concerning videos 163
15. Every Page is Page One Topics and the Big Picture 167
Books and the big picture 168
The priority of the big picture 169
Writing the big-picture topic 170
Finding the end of the string 172
Pathfinder topics 173
16. Sequence of Tasks vs. Sequence of Topics 177
Working backwards 179
17. EPPO and Minimalism 181
EPPO as a platform for minimalism 181
Is EPPO minimalist? 184
Minimal vs. comprehensive 185
18. Structured Writing 189
The varieties of structured writing 190
Benefits of computably structured writing 200
Structured writing and bottom-up organization 205
19. Metadata 207
The meaning of metadata 208
Topics should merit their metadata 209
Metadata comes first 214
20. Linking 215
Crowdsourced links 215

Soft linking based on subject affinities ... 215
Soft linking and list generation ... 219
21. Reuse ... 221
Reuse on the Web ... 221
Static vs. dynamic reuse ... 224
Other forms of reuse ... 225
Reuse, linking, and interactive pages ... 226
22. Making the Case for Every Page is Page One ... 227
EPPO and resource constraints ... 228
EPPO and continuous delivery ... 229
EPPO and content change ... 230
EPPO and content aging ... 231
EPPO and agile methodologies ... 233
EPPO and content management ... 235
EPPO and PDF/help ... 237
EPPO and content marketing ... 241
EPPO and DITA ... 246
EPPO and wikis ... 248
Making the case for technical communication on the Web ... 249
23. Afterword: EPPO, but Not for Everything ... 253
Glossary ... 257
Bibliography ... 263
Index ... 267

Foreword

DTRITTRPATRTITRLAFOTDOTCC is probably the longest acronym you've ever run into. While it doesn't roll off the tongue easily, and it's difficult to memorize, it is the acronym that most represents the intent of the modern technical communicator:

> To deliver the right information to the right person at the right time in the right language and format on the device of the customer's choosing.

How to go about accomplishing this goal has been the focus of technical communication and content strategy thought leaders around the globe. Ann Rockley, Rahel Bailie, Sarah O'Keefe, Robert Glushko, JoAnn Hackos, Joe Gollner and others have spent considerable time thinking through the challenges involved in accomplishing this lofty goal. They (and others in our field) have developed, tested, and implemented methods, standards, and tools designed especially for tackling this challenge. And, they've willingly shared best practices and lessons learned discovered along the way.

Out of this body of knowledge came important innovations – single-sourcing, multi-channel publishing, component content management, and structured authoring – process improvements that resulted in the elimination of unnecessary manual tasks, the automation of tasks best performed by computers, and tremendous savings in content destined for translation.

While there is no doubt that these efforts created tremendous value for the organizations we serve, did these improvements help us create better content?

In *Every Page Is Page One*, Mark Baker argues that our focus on serving the needs of the organization has done little to improve the usefulness of our content. Single-sourced, multi-channel, XML publishing projects can indeed help organizations save money (compared to less efficient production methods), but these methods don't do much to improve the value of the content to those who matter most: our audience.

While not everyone will agree with his thinking, Baker makes valid points that deserve to become part of our professional discourse. For instance, why do documentation projects continue to follow outdated publishing models? Are DITA topics appropriate for the web? Are they really self-contained information modules? And, assuming they are, do they provide the context required of people who stumble across them while foraging for information in a web browser, perhaps on a tablet or a smartphone?

Part manifesto, part textbook, *Every Page Is Page One* should be required reading for all technical communication professionals. Not only is this book loaded with thought-provoking ideas about how we might increase the usefulness of the content we create, it's also loaded with information that software and services vendors need in order to create tools that help technical communicators DTRITTRPATRTITRLAFOTDOTCC.

My advice: Read this book and take its lessons into account when you create your next documentation project. If you do, chances are you'll improve the utility of your content and dazzle your audience. Who knows, you might even create content that your customers find useful – content they might actually want to read.

Scott Abel
The Content Wrangler
October 2, 2013

Preface: In the Context of the Web

There is a scene in the James Bond movie Skyfall in which Bond approaches the bad guy's lair, an island covered by spooky abandoned tenements. It is a striking image, so while I was watching the scene, I pulled out my phone to see if it's a real place. It is. It is Hashima Island in Japan, an abandoned mining town. I'm not alone. Going online while watching TV is something most people do these days.[1]

In addition, I use Google Maps to follow the hero's journey when I am reading fiction and to check assertions and follow tangents when I am reading non-fiction. Even when the content I am consuming is not on the Web, I am.

This is highly important because it means that there is really no such thing as off-line content anymore. Even if the content is not found online, it is consumed online because the reader is online. This means that the way readers consume content online is now the way they consume all content, because they are always online. All content is consumed in the context of the Web.

Gerry McGovern reckons that the very idea of "going online" is outdated. We simply are online, all the time. "The Internet has become so pervasive that people don't think they are on it anymore, even when they are."[2] Today we live, work, and read in the context of the Web.

The way readers consume content online is now the way they consume all content, because they are always online.

When I was growing up, I had access to a town library, a bookstore at the mall, and a university library. I thought I was living in an age of information abundance. Today I realize that I was living in a age of information scarcity that was, in terms of practical access to information, closer to the middle ages than to the modern world. We live in an age of cheap and abundant information. Of course, abundance is not the same as quality, but abundance on this scale profoundly changes the culture and economics of information.

[1] http://www.huffingtonpost.com/2013/04/09/tv-multitasking_n_3040012.html

[2] http://www.gerrymcgovern.com/new-thinking/there-no-such-thing-internet-or-web-anymore

Technical communicators must adapt to these changes as they design, organize, and deliver content. They can no longer create help systems and manuals as they have in the past; customer expectations have changed too much.

Even for documentation that is not on the Web, all the recent customer feedback data I have looked at indicates that users do not think in terms of individual manuals and references. They only think of "the documentation," and they expect to be able to search and navigate it as one resource. Even if the documentation is not on the Web, readers expect every search to work like Google and every documentation set to work like the Web.

Technical communicators cannot create help systems and manuals the way they did before the Web.

I call this change of expectations: **Every Page is Page One**. People have multiple sources of information available all the time, and they hop freely from one to another. Authors don't dictate the reading order, readers do. And with every hop, readers arrive at a new page one.

Every Page is Page One is both an information design pattern and a content navigation pattern. For readers who live and work in the context of the web, Every Page is Page One is the dominant mode for finding and using information. Even if your content is not (or not yet) on the Web, you and your readers are best served by content that is written and organized for this new reality.

In this book, I discuss how the Web has changed the way people find and use information, how to adapt to these changes, and how to create content that is usable and navigable in an environment where Every Page is Page One.

Audience

This book is for technical writers, information architects, content strategists, and anyone interested in designing information that will be consumed on the Web or in the context of the Web. Even if you produce manuals and help systems, your users now consume your content in the context of the Web, with beliefs and expectations formed on the Web. This book is for you too.

Every Page is Page One is an information design pattern, not a technology. You can create Every Page is Page One content in any medium and with any authoring tool. Though certain kinds of tools can definitely help, Every Page is Page One does not require a tool change. Whether you work in DITA, FrameMaker, Word, a wiki, a Web CMS, or with pen and paper, this book is for you.

Discussion

Since well before I even thought of writing this book, I have been blogging about the Every Page is Page One concept at `EveryPageisPageOne.com`. Some of the material here originated on the blog, but this book is not a collection of blog posts. However, the blog is a good place to talk about the book and the concepts it champions, and I invite you to do so. The blog also contains a ongoing list of places where you can find Every Page is Page One content.

You can also discuss the book on the Every Page is Page One group[4] on LinkedIn.

Acknowledgments

Many smart and kind people, including many leaders in the field, have influenced this book, both directly and indirectly. A number of them graciously provided comments and suggestions that led in many fascinating directions, not all of which I have had the time, or space, or wit to follow. Therefore to all who read I say, this book is not, for me, a destination but a milestone. There is further, much further, to go.

Thanks are due to:

My wife, Anna, for saying "stop talking about it and just get on with it."

My publisher, Richard Hamilton, for saying "I'd like to publish it," and a great many other useful things besides.

Everyone who has commented on my blog, particularly those who have held my feet to the fire and forced me to really think through and properly support what I was

---

[4] http://www.linkedin.com/groups/Every-Page-is-Page-One-4671518

trying to say. To name a few: Scott Abel, Alan Brandon, Frank Buffum, Pamela Clark, Ray Gallon, Vinish Garg, Anne Gentle, Joe Gollner, Yuriy Guskov, Alan Houser, Steve Janoff, Tom Johnson, Marcia Johnston, Neal Kaplan, Alex Knappe, Larry Kunz, Jonatan Lundin, Gordon McLean, Paul Monk, Joe Pairman, Tim Penner, Myron Porter, Ellis Pratt, Ann Rockley, Barbara Saunders, Dan Schulte, David Singer, Val Swisher, Kai Weber, Leigh White, and David Worsick.

My fellow tech comm and content strategy bloggers, whose work has inspired, provoked, and informed me as I worked out the ideas in this book: Laura Creekmore, David Farbey, Ray Gallon, Joe Gollner, Tom Johnson, Larry Kunz, Gordon McLean, Sarah O'Keefe, Ellis Pratt, Alan Pringle, Val Swisher, Julio Vazquez, Kai Weber, and Leigh White.

The very generous people who took the time to review the book in the proof stage and provide frank and pointed feedback. It will be very evident to them when they see the final book just how profound their influence on its shape and argument has been. They are: Helen Abbott, Pamela Clark, Ray Gallon, JoAnn Hackos, Alan Houser, Tom Johnson, Larry Kunz, Jonatan Lundin, Joe Pairman, Ellis Pratt, Val Swisher, Sara Wachter-Boettcher, Tina Klein Walsh, Kai Weber, and David Weinberger.

The many colleagues and collaborators whose influence and encouragement all in some way contributed to this book, including: Roy Amodeo, Helen Arrowood, Christy Morton Bhatnagar, Pamela Clark, Carla Corcoran, Leona Gray, Jennifer Keene-More, Carol Miksik, Bill Petrie, Cindy Sprague, Tina Klein Walsh, Sam Wilmott, Norbert Winklareth, Ron Zwierzchowski, and, particularly, Christopher Gales for having faith.

CHAPTER 1

Introduction

Studies by Peter Pirolli, Stuart Card, Kim Chen, and Ed H. Chi of PARK[9] show that people's behavior on the Web follows a pattern similar to the optimal foraging patterns of wild animals. The name for this behavior is *information foraging*. Just as wild animals follow patterns that allow them to find adequate nutrition with the minimum expenditure of calories, information seekers follow patterns that allow them to find adequate information with the minimum expenditure of mental energy.

The key concept in information foraging is information scent. Just as an animal follows its nose, so an information seeker follows the scent of information. And just as an animal will move on to a different foraging ground when the smell of food grows weak, the information forager will move on to a different source when the scent of information grows weak.[1]

In other words, people do not search for information with the intellect of a research librarian, but with the nose of a predator. We look for the patches of content that our nose tells us are most likely to yield the information we are after.

People do not search for information with the intellect of a research librarian, but with the nose of a predator.

In his Alertbox article "Information Foraging: Why Google Makes People Leave Your Site Faster"[21], Jakob Nielsen describes the kind of behavior that results from following our foraging instincts:

> A fox lives in a forest with two kinds of rabbits: big ones and small ones. Which should it eat? The answer is not always "the big rabbits."
>
> Whether to eat big or small depends on how easy a rabbit is to catch. If big rabbits are very difficult to catch, the fox is better off letting them go and concentrating exclusively on hunting and eating small ones. If the fox sees a big rabbit, it should let it pass: the probability of a catch is too low to justify the energy consumed by the hunt.

[1] The Wikipedia article [http://en.wikipedia.org/wiki/Information_foraging] provides a good summary and links to the research if you are interested.

This foraging behavior is not exclusive to the Web, of course. John Carroll saw the same behavior in his research subjects using paper manuals.

> Learners also often skip over crucial material if it does not address their current task-oriented concerns or skip around among several manuals, composing their own ersatz instructional procedures on the fly.
>
> —*The Nurnberg Funnel*[8, p. 8]

But while information foraging is not unique to the Web, the Web profoundly changes foraging patterns. In the paper world, a book is an information patch, but it is quite expensive to move to a different patch, so information foragers are motivated to stay in their current patch and hunt it out, getting every calorie of information before moving on. In the context of the Web, moving from one information patch to another costs almost nothing. Therefore, the optimal strategy for an information forager is not to hunt out one patch, expending more energy on scarcer and scarcer game, but to move on to a fresh patch with tastier, more plentiful game.

Nielsen explains how the power of Google and ubiquitous connectivity affect information seeking behavior[21].

> Information foraging predicts that the easier it is to find good patches, the quicker users will leave a patch. Thus, the better search engines get at highlighting quality sites, the **less time users will spend on any one site**.
>
> [A]lways-on connections encourage **information snacking**, where users go online briefly, looking for quick answers. The upside is that users will **visit more frequently**, since they have more sessions, will find you more often, and will leave other sites faster.

The richer the information environment, the more widely the information forager will range. This forces a change in strategy for content that you want to be found and consumed. As Nielsen puts it:

> The two main strategies are to make your content look like a **nutritious meal** and signal that it's an **easy catch**. These strategies must be used in combination: users will leave if the content is good but hard to find, or if it's easy to find but offers only empty calories[21].

And because even offline content is now consumed in the context of the Web, by readers who are online even if the content is not, it is the Web's foraging conditions that determine how long a reader will stay in that content.

The Web is an almost perfect information foraging environment. It is full of small morsels that are easy to catch and easy to chew. This encourages the type of quick information snacking behavior Nielsen describes. It puts a premium on the short and easily obtainable, and it changes information consumption habits beyond the Web. Readers have become habituated to information snacking.

Every page is page one

The consequence of this information snacking behavior is that every page the reader reads becomes a new page one. When you search for information on the Web, whether you use a search engine or follow a link, and you land on one of the billions of pages on the Web, that page, for you, is page one.

This is simply the way the Web works. There is no "Start Here" page for the Web. Wherever you dip your toe into the Web, that is your page one. We can't avoid this. Whether you are a reader or a writer, and whether you like it or not, that is the way the Web works. Every page is page one.

On the Web, every page is page one.

Of course, not every page on the Web makes a good page one. Many pages do work as page one, but a distressingly large number do not. And many of the pages that don't work were produced by professional writers working for established companies.

The professional writer carefully plans and constructs an ordered set of content, and more often than not, the order is a hierarchy or sequence, as in a book. Only one page is page one. The other pages descend from and rely upon page one and all the pages that stand between them and page one.

Authors who have been trained to write books construct page 16, page 187, or page 2596. While their page 187 may be a brilliantly conceived and executed page 187, it probably doesn't work as page one for a reader who lands there from a search or, for that matter, from anywhere. Content consumed on the Web or in the context of the Web is increasingly consumed the same way: as if every page is page one.

The Web can be thought of as a giant noticeboard on which you pin individual disconnected pages in the hope that someone will notice them. But the Web is much more. It is a hypertext medium consisting of a navigable network of interconnected pages. The places where content works on the Web are neither book-like things forced online nor random pages posted carelessly. The best are integrated, highly navigable collections of Every Page is Page One pages created by people who understand the Web. (This group includes many younger professional writers who have never written for any medium other than the Web.)

What is needed today is the same rigor and discipline professional writers have long brought to making books, but not the same methodology. The book model does not work for the Web or for content consumed in the context of the Web. My aim in writing this book is to begin to define a rigor and a discipline for writing and organizing Every Page is Page One pages.

> NOTE:
>
> Because the word *page* refers to a mechanical division of content, which is not, either on paper or on the Web, necessarily a logical unit of meaning, I will use the term *topic* rather than *page*, unless I am referring literally to a page.

Why do we still write books?

Why, after all this time, do so many tech writers still produce books and organize help systems as if they were books? It's not as if we have any illusion that people read them. The expression *RTFM* was already shopworn when I entered the profession more than 20 years ago. John Carroll's research[8] showed that adult learners learn by exploring, not reading.

> People seem more interested in action, in working on real tasks, than in reading. We found that learners were given to plunging into a procedure as soon as it was mentioned or of trying to execute purely expository descriptions.
>
> —John Carroll, *The Nurnberg Funnel*[8, p. 8]

Users dive into a product, work till they get stuck, and then look for quick answers to get them unstuck. Of course, it's not literally true that no one reads the manual. In some cases, there is no choice, and in some cases, it is the only place to turn when you get stuck. But few relish the experience.

Because we know that people don't read manuals like books, we stuff them full of indexes, subheadings, tables, and other eye-catching and search-enabling devices.

All this was well known before the Internet made Google junkies of us all. Today, of course, people who need a bit of technical information search for it. They don't sit down and read technical manuals cover to cover. As David Weinberger has pointed out, the power to organize information has passed from the writer to the reader (*Everything Is Miscellaneous: The Power of the New Digital Disorder*[28, pp. 22–23]). Yet still we write books. Even when we adopt topic-based tools, we often use them to build books. Why?

In their book *Switch*[14], Chip and Dan Heath cite the research of James March, which holds that people make decisions based on one of two models: the consequences model or the identity model. Decisions based on the consequences model are made by looking at what will happen as a result of the decision: "if I throw this brick through this jeweler's window, I will get arrested and go to jail." Decisions based on the identity model are made by asking what kind of person I am: "although there are no cops around, I am not the sort of person who throws bricks through jeweler's windows."

Since we have known for decades that people don't read the books we write, it is hard to imagine that we continue to write them based on the consequences model. We don't write books because we think they are the best way to communicate technical information. We know they aren't. We have known for years. If March's research is correct (and once stated, it seems obviously true) then the decision to write books, despite their being so ineffective, can only be the result of the identity model at work.

We write books because we are the sort of people who write books.

We write books because we are the sort of people who write books. But, even more so than in the past, our readers are not the sort of people who read books. At least, they aren't sort of people who turn to books to solve everyday practical problems. They may still turn to books for fiction or philosophy or professional development, but not to solve problems. For that they turn to the Web.

Our task, then, is to learn to write the kind of material that gets filtered in. We have to learn to write Every Page is Page One topics. More profoundly, we have to change how we validate our identity. We have to start thinking of ourselves as the kind of people who write Every Page is Page One topics.

This is, of course, a work in progress. The Web is still new, and we have by no means fully adapted to the Web. We can't say with any certainty what the Web will look like a decade from now nor what the role of books may be down the road. But the change is upon us, and it is moving very fast. We must do our best to keep up.

About the book

The book is divided into three parts.

- **Content in the context of the Web:** The Web has changed the way we communicate and the way we discover and share knowledge. Its influence on how we exchange information is far more profound than simply providing a new publishing platform. The topic—a short, functionally complete piece of information that is richly linked to other topics—is the natural information format of the Web.
- **Characteristics of Every Page is Page One Topics:** Every Page is Page One topics are the natural information form of the Web, but to learn to write them successfully, it is helpful to have a list of the major characteristics of good Every Page is Page One topics. This section outlines those characteristics.
- **Writing Every Page is Page Topics:** Writing Every Page is Page One topics for isolated subjects is one thing, but to document a complex product requires both discipline and preparation. This section looks at how to use the characteristics of Every Page is Page One topics to guide your writing, and how to handle the big picture and tasks with a strong sequential component to them. It also looks at tool choices and how to manage an EPPO development project.

Content in the Context of the Web

The Web is a hypertext medium. It has a million paths, but no starting place. In such an environment, every page is naturally page one. But why would anyone prefer to seek information in something so large and unruly as the Web, when you have provided a nice, orderly manual? This part looks at how the Web has changed the way people seek information and what that means for content creators.

CHAPTER 2

Include it all. Filter it afterward.

When readers forage the Web, rather than picking up a book, they are expressing a preference to, in David Weinberger's words from *Too Big to Know*[27, p. 176], "Include it all. Filter it afterward." Why does someone choose to forage the Web rather than open your manual or help system?

In the paper age, if you wanted a recipe for an omelet, you looked in a cookbook. You browsed your shelves for a cookbook and then searched the book for omelet recipes.

In the Web age, if you want an omelet recipe, you search the Web for "Omelet Recipes." Your source is the entire Web, which contains information on astronomy, psychiatry, celebrities, pornography, programming, classic cars, elephants, aliens, conspiracies, cover ups, presidents, paupers, photographers, blackmailers, fiction writing, oil painting, fixing flat tires, flying kites, and just about everything else, real and imaginary, that you can think of. You are searching everything.

You then apply a filter in the form of your search term "Omelet Recipes." Astonishingly, considering all the other stuff there is on the Web, your search engine – usually, but not always, Google – instantly provides you with a list of omelet recipes from around the Web and adroitly filters out millions of unrelated pages.

This is so commonplace that we don't stop to marvel. Yet it really is extraordinary that this works so well. And because it works so well, you don't need to think about what sources to search. Instead, you ask your question of the entire Web. Foraging the entire Web is now less effort than foraging for, and then in, a single book.

This profoundly changes what it means to provide a source of information since, most of the time, people no longer use individual sources of information. They include everything and filter it afterwards. Therefore, content that is not on the Web is much less likely to be found and consumed. And even when people are reading a book or viewing a help system, the Web is readily available. The content may not be on line, but the reader is.

Foraging the entire Web is easier than foraging for, and then in, a single book.

Just Google it

The most obvious expression of users' growing preference to filter for themselves is Google. In 2011, people searched Google, on average, 4,717,000,000 times per day[25].

Many writers are in denial about the power of Web search. There are too many false hits, they complain, too much stuff to wade through. It takes too long to find things. It's much easier to find things in a book with a well-prepared index.

The problem with this critique is that it assumes you already have the right book, which is seldom true. If you are sitting in your office with a shelf full of books behind you, the equivalent of performing a Web search is to turn around and face the bookshelf. At that point, you have a list of titles, most of which are irrelevant to your inquiry. Google also offers you a lot of choices, but it ranks them in order of likely relevance to your current inquiry; your bookshelf does not do that.

Perhaps you already know which book contains the answer you need. But you still have to get up, cross the room, pull the book off the shelf, and consult the index to find the page you want. By contrast, Google not only finds the site that contains the information you want, it provides a link directly to the page in question. You don't have to leave your chair or take your eyes off the screen. You can evaluate multiple search results in the time it takes to get one book off the shelf and navigate the index.

And this assumes that you are looking in a well-indexed book that you already own. (How many of the books you own are well indexed?) It also assumes that you pick the correct book the first time. If you have to check several books, especially if they don't all have good indexes, all that searching will take far longer than sifting through search results.

What if you discover that the information you want is not in any of the books you own? Then it is off to the bookstore or the library to hunt through the stacks, the card catalog, and the indexes (where available) of dozens of potential books; stand in line to check them out or buy them; and then carry them home. In that time you could have done dozens of searches and evaluated hundreds of results.

You could also have posted your question on a forum and almost certainly received multiple answers before your paper-world equivalent returned from the library. For example, the median time to get an answer to a software development question on Stack Overflow[1] – a popular forum for programmers – is 11 minutes[18].

In other words, even in the paper world where writers, publishers, and librarians ostensibly do the work of assembling and filtering content for you, you still have to do a great deal of leg work (literally) to get to the content you need.

The median time to get an answer to a question on Stack Overflow is 11 minutes.

Ebooks change this picture somewhat, cutting out the physical journey to the shelf or the library, but you still have to find the right ebook before you can search it and finding the right book can still be a lengthy and, in some cases, costly process. In fact, the best way to find the right ebook is probably to search the Web for it.

People prefer to search the Web because it lets them find more content in less time.

The long tail

Books are relatively expensive to produce and distribute. By contrast, putting information on the Web costs little more than the time it takes to type it. This means that all kinds of information gets put on the Web that would never find its way into a book. This doesn't mean the information is necessarily of lower quality, just that there is less aggregate demand for it.

The amount of low-aggregate-demand content on the Web is enormous. In his book *The Long Tail: Why the Future of Business Is Selling Less of More*[1], Chris Anderson shows how the Web changes the market for low demand items by making them more easily available. A *long tail* is a statistical distribution in which an unusually high number of occurrences appear far from the center of the distribution. In other words, you get a distribution that looks more like an L shape, with more items away from the center than in a normal distribution (see Figure 2.1).

[1] http://stackoverflow.com

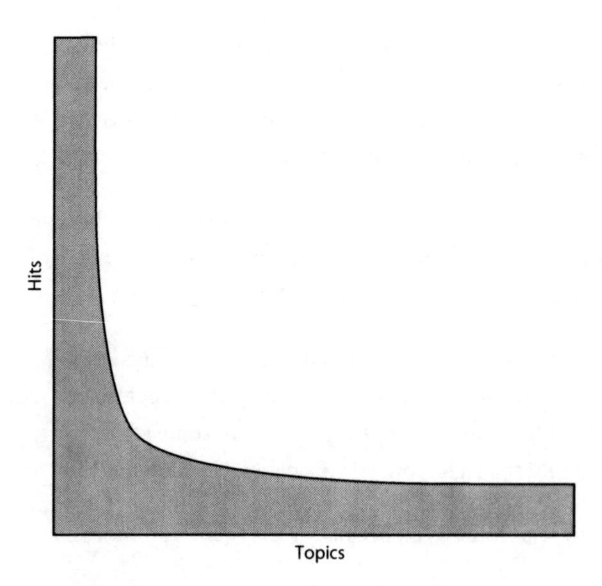

Figure 2.1 – The long tail distribution

Think about your local supermarket. It stocks the staples that everyone buys: bread, milk, bananas, and so forth. But it also stocks some pretty obscure things. Somewhere in aisle twelve on the fourth shelf of the third section there is a single row of small jars containing something truly exotic, and in the produce department you may find a fruit you've never seen before.

Why does the store stock all these obscure items? Because they recognize that the few people who want those little jars or the unusual fruit also want bread, milk, and bananas. If the people who want these obscure items were a distinct set from the people who want bread, milk and bananas, the store would probably drop those items to free up space. Instead, they stock these items because they know almost all of their customers want one or more of them in addition to their staples (Figure 2.2).

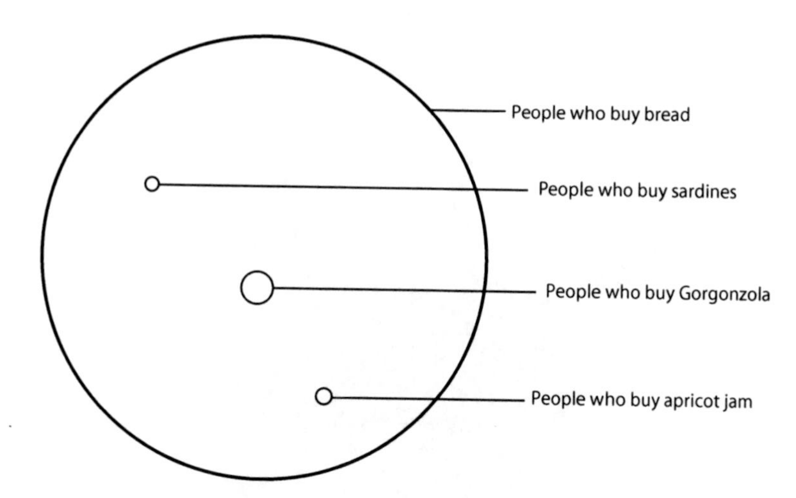

Figure 2.2 – Buying bread and Gorgonzola

Suppose Dave wants bread and sardines, June wants bread and Gorgonzola, and Pete wants bread and apricot jam. Bread is the product in high demand, but Dave, June, and Pete probably won't shop at a store that stocks only bread (unless the bread is extraordinary). They will shop where they can get everything they need in one store. If you want all of a person's business, you must meet all of that person's needs, even the obscure ones.

Of course, it isn't usually possible to write a document that covers all of your customer's needs. Even if you identify and provide all the high-demand content, each individual user will want low-demand content some of the time. As Figure 2.3 shows, the content in high demand is at the intersection of the demands of many users, each of whom also has content needs that are outside the high-demand zone. If your documentation provides only high-demand content – which is generally all most organizations have the resources to provide – every user will be disappointed at least some of the time. Search results may not be as neatly organized as your document, but your reader's chances of finding an answer are much higher.

If you want all of a person's business, you must meet all of that person's needs, even the obscure ones.

Figure 2.3 – Coverage of high-demand content

Do information needs follow the long tail distribution? Yes, in fact it was in the field of information delivery that the commercial importance of the long tail came to light. A study by Brynjolfsson, Hu, and Smith[5] showed that Amazon was realizing an increasing volume of sales from obscure books and that as access to the long tail got easier, demand in the long tail picked up. In another paper, the same authors found that "Amazon's Long Tail has gotten significantly longer from 2000 to 2008 and that overall consumer surplus gains from product variety at Amazon increased five-fold from 2000 to 2008"[6].

We can see the long tail everywhere on the Web. In product forums, people ask obscure questions and get answers. And as more long tail questions get answered, more people post questions and more of them get answered. The growth of long tail information drives demand for information in the long tail.

So how often does documentation disappoint users? I think it's more than half the time. I suspect that even if you document all the high-demand material perfectly, you will only cover half of what users are looking for. And I suspect that percentage feels even worse than it is. Not finding something takes much longer than finding it – you have to exhaust many possibilities before you can give up and conclude that the information just isn't there. Even if you succeed half the time, it will feel like you fail most of the time because you will spend most of your time on searches that fail.

The Web may not cover the high-demand content as well as your docs, but it tends to cover the long tail pretty well. Even though a bakery may make better bread, most people buy their bread from the supermarket because they can pick up their sardines, Gorgonzola, and apricot jam at the same time.

Of course, the Web does not meet all information needs. One of my most-thumbed reference works is Michael Kay's *XSLT Programmer's Guide.* While there are many XSLT references on the Web, none has the depth of Kay's book. If I have a hard XSLT question to solve, his book is my best resource. But it isn't the first place I look. I turn to it only when search fails. A search may not turn up as thorough a reference, but it will probably return more applicable code snippets. I wish the information in Kay's book was available on the Web, because the Web is still the first place I look.

By the same token, if your documentation contains information that is not available on the Web, your users will be frustrated by a Web search. They may find the long tail but not the dog. And if they are disappointed in their web search, it's just as bad as if they were disappointed in your docs – either way, your product will be seen as hard to use. For your documentation to add value, it must be available wherever people look for information about your product, and they will not just look at the information you provide, they will look at all the information that other people provide as well.

If users are disappointed in their web search, it's just as bad as if they were disappointed in your docs.

If you provide high-quality, high-demand content on the Web, Google and other search engines will find it and bring it to the top of the results when your users search for your information. But search engines also return results from the long tail – all of the low-demand content you could never possibly hope to provide, but which, taken together, will supply a significant portion your users' information needs. Because your content will appear as just another search result, you need to make sure it's content that users filter into the results they choose to read. If the search lands users in the middle of some vast book, they will probably be frustrated and filter your content out. To be filtered in, your content needs to work as page one.

You may think the long tail does not apply to you because your product has a small user community and there won't be much content about it on the Web. If you think this, do a web search with questions about your product. You may be surprised at what you find. But even if there is not (yet) a long tail of content for your product, your users have no way of knowing that. They will search the Web because that works most of the time. Don't expect your users to change their information seeking habits just for your documentation.

Don't expect your users to change their information seeking habits just for your documentation.

Also, if a Web search for your information fails, your users may conclude there is no manual. Therefore, if there is no long tail information for your product, it is actually doubly important to get your content onto the Web.

Of course, Google doesn't have access to all the world's knowledge, and its relevance algorithms are better for some queries than for others. No single resource, Web-based or not, has access to all the world's knowledge, and while a web search may be a good starting place, some forms of research require more than a Web search can provide.

But such research is expensive, requires a high level of research skills, and may cost more than would be gained by solving your problem. In practice, people's behavior is *satisficing*. That is, people will generally choose something that is good enough and easy to get over something that is superior but requires more work. As described in Wikipedia,

> In decision making, satisficing explains the tendency to select the first option that meets a given need or select the option that seems to address most needs rather than the "optimal" solution.[2]

In simple terms, what Google provides is often good enough.

---

[2] http://en.wikipedia.org/wiki/Satisficing

However, it doesn't really matter how good any search engine is or how good the information on the Web is. What matters is that the Web has, in Malcolm Gladwell's term, tipped.[3] The Web is now the place where people go for information.

What Google finds is often good enough.

Of course, Web searches sometimes fail. And the searcher will not know why. It could be poor search technique. It could be a defect in the relevance algorithm. It could be that there simply is no information on that subject available on the Web. The searcher can't tell. The search engine can't tell either. The cost of finding out is high, and the skills required are specialized. Therefore, if several search attempts return no results, most people will accept an empty result and move on with their lives. If they are seeking help on a product, they might try that product's help system, if they still remember about help systems, or the manual, if they still remember where they put it. Or they may not. It depends on how much effort they are willing to expend on finding an answer. Sometimes satisficing behavior means taking no for an answer.

Authority and experience

Recently, I awoke to find my inbox full of moderation emails for my blog. Sometime during the night, the spam filter lost the ability to connect to its server and started dumping all the spam comments into the moderation queue. Not only did I need to deal with the backlog, but new messages kept popping up every few minutes.

The first thing I did was to check that WordPress and my themes and plugins were up to date. I discovered that two plugins and the theme were out of date, but when I tried to update them, WordPress reported that it could not reach the server. I searched Google for the problem and found a post on the WordPress forum that suggested switching to a default theme, doing the upgrades, and then switching back to the normal theme.

This sounded a little improbable, but as I kept reading, I found posts from people who had tried this remedy with success. So I tried it and the trick worked.

[3] *The Tipping Point*[13]

I don't know if anyone at Automattic (the makers of WordPress) knows anything about this, has documented it, or scheduled it to be fixed. If they have, my Google search did not turn up anything about it. But I found a fellow blogger who had the same problem, found a solution, and documented it. And I found other bloggers who confirmed that the solution worked for them. I never found the voice of authority, but I found the voice of experience.

I never found the voice of authority, but I found the voice of experience.

One of the ways the Web changes how we think about knowledge and information is that it give us access to people whose expertise comes from experience as well as those with credentials. In the words of Beth Noveck, director of President Obama's Open Government Initiative (as quoted by David Weinberger):

> Do we think about expertise as experiential or academic? Books or context-centered? The person driving a truck every day or the logistics expert from IT? The answer is both, of course. Now the technology lets you find experienced people as easily as credentialed ones.
>
> —David Weinberger, *Too big to know*[27, p. 72]

This is a profound change. In the past, your access to experience was local. If you didn't know someone close to you who had done the task before, you had to fall back on reading the book. But today, you can type your query into Google (or YouTube, or StackOverflow) and get immediate access to people who have successfully completed the exact task you want to do.

Today, you can type your query into Google and immediately get access to people who've successfully completed the task you want to do.

Chances are you will find the record of a conversation between someone who has done the task and someone who wants to do it. Usually the conversation is in a forum, a mailing list archive, or a technical exchange portal like Stack Overflow or Quora. If you don't find an exact match, you can usually find the best place to ask your question.

Real-world experience is something you generally can't get from the documentation, no matter how thorough or well organized it is. The procedure may be there and it may have been tested in a lab, but the voice of experience – the person who has actually done this in the real world – is not there. Thus, users often lack confidence in the procedure described in a book, either because they are not sure they understand the procedure correctly or because they aren't sure they are in the correct context.

Speaking, even virtually, to someone who has done the task in real life helps build confidence. And with YouTube and other video sites, you can often find a video showing you exactly how someone has performed this task.

We often assume customers regard the company documentation as containing the most trustworthy information. However, trust is fundamentally something we bestow on people, not books or companies. In many cases, members of a user community trust each other far more than they trust the vendor, who they know has commercial interests which may differ from their own. Trust is not conveyed by the gilt letters on the spine of a book anymore. People trust the people who have helped them in the past and those who have helped others like them. You can be that person for them, but you can't assume that trust is given automatically.

In her book, *Conversation and Community*[12], Anne Gentle describes how the Web has enabled groups of individuals with like interests to come together and share content directly with each other.

> The harder it is to get something to work, the more comforting it is to get help from a real person. The increased complexity and reach of technology has driven us to reach out to people through various methods to help us understand and use that technology. How has changing technology affected how these conversations are held?
>
> An example of how technical documentation has changed over the years is sewing patterns. Historically, people talked face-to-face to teach others how, for example, to sew a quilt for a bed cover. Later, the patterns were written down, and with the advent of printing, patterns could be produced for anyone to use at home. Today's quilting instructions are disseminated on the Internet, mailing lists, forums, and blogs, and online communities are forming around the passion for quilts and quilting.
>
> —Anne Gentle, *Conversation and Community*[12]

People prefer to search the Web because it gives them access to experience as well as credentials.

This preference for conversation, community, and the voice of experience is not universal. Some cultures rank authority much more highly than others, and some readers will rank information more highly if it comes from the manufacturer. However, even these readers don't always turn to the manual before going to the Web. They may still prefer to include it all and filter it afterwards; they just tune their filter a little differently. So even if some users prefer information from the manufacturer, that information still has to be available to be filtered in.

Aggregation and curation

The Web is not just an environment for searching, it is also an environment in which you can aggregate and curate content and tailor collections of content for you own needs. In her book *Content Everywhere*[26], Sara Wachter-Boettcher writes:

> Content shifting ... is the process of taking content from one context – as in, a specific website or application – and shifting it to another location. People shift content for a wide range of reasons. They may want a cleaner interface for reading, seek to save the content for later, include the content in a collection of similar pieces, or need to access it at a time when they won't have an Internet connection. Whatever the reason, though, people are shifting content more and more often – and tools are cropping up left and right to help them do it.
>
> ...
>
> This is the world of reusable and reconfigurable content: content that can be pushed out lots of places at once, assembled and associated with other relevant bits on the fly, displayed in different combinations for different purposes, or connected and combined by users themselves.[26]

Aggregation and curation is more than finding existing things and pinning them somewhere. It includes subscribing to content feeds on a chosen subject so you can received the latest information as soon as it is created.

> This approach to personalized content, where a user can select the information she needs and create her own unique set of dashboards, reports, or other content collections, is becoming increasingly important in areas like healthcare, government, and education – places where users require different information depending on their needs, and where they tend to return regularly for news and updates.[26]

People prefer to search the Web because the Web lets them aggregate and curate content dynamically.

Filter it afterward

There are many reasons why readers prefer to include everything and filter it afterward. But how do they do the filtering, and where are the filtering tools? While you can look at the Web as a vast chaotic collection of miscellaneous information (*Everything is Miscellaneous*, is how Weinberger describes it), you can equally think of it as a vast information filter.

The Web is full of filters, and most of us spend most of our time on the Web creating and using those filters. Consider the following:

- **Google is a filter:** A search may return thousands of results, but Google has already filtered out billions of pages. Then it orders the results using a relevance algorithm, which is a second filter that tries to bring the most relevant results to the top. Sure, there may be some irrelevant hits and some searches may be confused by different meanings of the search terms, but it is astonishing how often Google returns highly relevant content on the first page. The current generation of search engines, of which Google is the best-known example, are extraordinarily powerful filters. Without these filters, the Web would not work.
- **Twitter is a filter:** Millions of tweets are sent everyday. You filter that massive stream by selecting who you will follow and which hash tags you will watch. Your whole experience on Twitter is driven by the filters you create. Twitter also helps you filter the wider Web. You choose people to follow whose interests are similar to yours, and they tweet links to content you might be interested in. I rely on

Without modern search engines, the Web would not work.

Twitter as a live filter that keeps me up to date on what is being thought and said in the worlds of #techcomm and #contentstrategy.

- **Facebook is a filter:** Facebook lets you filter both what you reveal to the world and what you see of what the world reveals to you. The filtering algorithm is so central to Facebook that it is a source of constant tinkering, criticism, and debate.
- **LinkedIn is a filter:** By choosing which people you link to and which groups you join, you control what part of LinkedIn content reaches your in-box. And LinkedIn provides sophisticated filters to help recruiters and sales people find prospective employees or customers.
- **Amazon is a filter:** Though it stocks just about every book on the planet (the long tail), Amazon succeeds in showing you a half dozen titles you might genuinely be interested in based on what you have bought and what others who bought the same thing have bought.
- **Reddit and Pinterest are filters.** Content aggregation and curation sites are filters; even the humble RSS feed is a filter.
- **YouTube is a filter:** It constantly (and effectively) suggests videos you might want to watch based on what you are watching now and have watched in the past.
- **Q&A sites are filters:** Chris Parnin, et al, discuss this in their paper about crowdsourced API documentation on Stack Overflow:

> Unlike traditional documentation, crowd documentation can be filtered using various quality attributes. Each thread on Stack Overflow has a number of quantitative properties, such as the number of views, the number of votes and the score based on those votes), whether it has been answered and whether the answer has been accepted and/or bountied, and how many times it has been favorited. While some of these curation filters are explicit (such as votes), others are implicit (such as views). Stack Overflow uses these filters to provide sorting of Q&A threads, e.g., threads on Stack Overflow can be sorted by votes and level of activity. To ensure a certain level of quality (e.g., only consider threads with an accepted answer), the crowd documentation can be filtered to only include threads that meet a certain threshold.[23]

- **Ad tracking:** Advertisers track the sites you visit and the things you buy, and then they use that information as a filter to select targeted ads to show you.
- **Web links:** Every web page that contains hyperlinks is a filter, selecting and pointing to content that may be of interest to people reading that page.

The Web works because it includes everything and because it provides the means to filter everything to get just what you need. You can give your readers additional filtering tools, but remember that they already use and are familiar with a wide variety of filtering tools. If your content is not accessible to those filters, your users may never see it. As David Weinberger notes:

> Old knowledge institutions like newspapers, encyclopedias, and textbooks got much of their authority from the fact that they filtered information for the rest of us. If our social networks are our new filters, then authority is shifting from experts in faraway offices to the network of people we know, like, and respect.[27, p. 10]

Therefore, the key question for authors is not how readers filter the content they consume. Rather, the question is how authors can make sure their content gets filtered in, not filtered out.

CHAPTER 3

The Distributed Nature of Content on the Web

A lot of the patterns we use for designing and managing online content, particularly in tech comm, come from what we learned in the book world. For example, many websites are structured as a tree, a standard pattern for books. The trunk is the homepage. You climb from there, upward and outward, until finally you reach the leaves. That is how we design websites. That is how we test websites.

But, in many cases, that's not how we use websites.

In many cases, in fact, we don't use websites at all. What we actually use are:

1. The Web
2. Web pages

How we use the Web

Users who set out to consult your docs, and your docs alone, may go straight to your site and look for the documentation link. But if, for any of the reasons we looked at in Chapter 2, they decide to do a Web search – to include it all and filter it afterward – the search results may or may not contain hits from your site, and there is a good chance searchers won't pay much attention to where the results came from.

Search engine results don't lead us to websites, they lead us to individual web pages. Those pages are, of course, part of a website, but we don't experience them as such. We experience them as individual pages.

If we find a particularly useful page, we bookmark the page, not the website. If we want to share with friends, we share the page, not the website.

Sometimes a good page will have links to other pages. If we visit them, we go straight from page to page. We do not navigate the website. The page might be on another website, but there is a chance we won't even notice.

After we have finished on a page, we may hit the back button to go back to the previous page or we may search for something else. But we do not navigate the website to get there. The previous page might be on another website, but there is a chance we won't even notice. If we are curious about something on a page, and there is no obvious link to a page that explains it, we highlight the phrase and search the web for it. We do not use the website, we use the Web.

Dynamic semantic clustering

When we publish to the Web, we tend to think readers will see our web content as one tight little cluster of content, all together in one place, as shown in Figure 3.1.

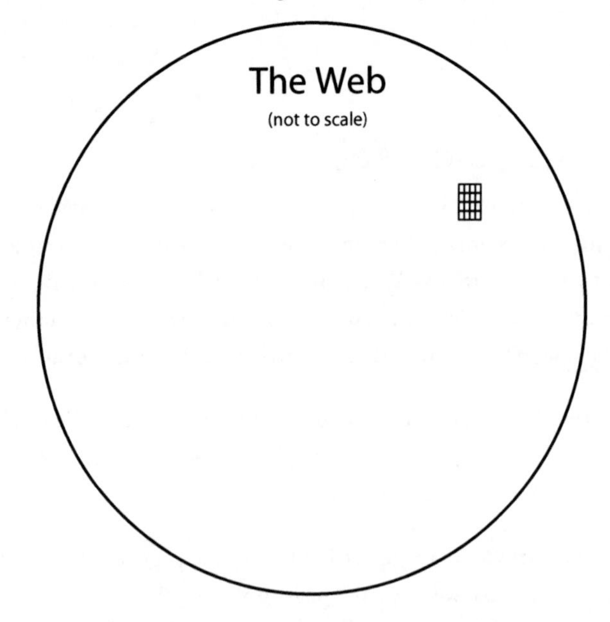

Figure 3.1 – How you think your content appears on the Web

After all, we created a website, and a website is a separate entity consisting of tightly associated content, isn't it? If people navigate to your home page, this is indeed how your content will appear. But when people navigate by search engines and socially curated links, that isn't how your pages appear.

When you do a search or apply a filter to the Web, you get a set of results that is built dynamically around the semantics of the filter you applied, clustering pages related to your search terms or filter. We can call this process *dynamic semantic clustering.*

Dynamic semantic clustering has the effect of scattering your pages across the Web. Your pages appear individually in many different searches on many different topics. The only thing that brings pages together in a search is their relevance to a particular search string. While all your pages may reside under the same domain name and adhere to a common template, to web searchers they are scattered all over the landscape, as shown in Figure 3.2.

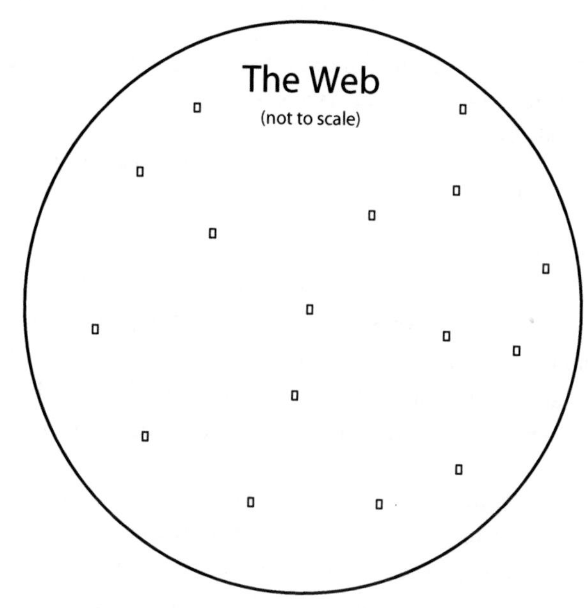

Figure 3.2 – How your content appears to people who search.

For example, even though every page on Wikipedia bears a common domain name (wikipedia.org), you probably don't experience it as a single website. Instead, you probably experience it as individual items in a thousand different Web searches that cluster a Wikipedia page on your subject of interest with other pages on that subject from other sites. As a search engine sees it, Wikipedia is scattered across the Web.

As a search engine sees it, Wikipedia is scattered all across the Web.

A common domain and a common style do not connect your pages to one another. Every page is page one. Every page is the hub of the user's current experience and is bound to other pages only by direct links or subject matter affinity and relevance as determined by Google or other search engines.

Think about how you typically end up in a Wikipedia article. It is usually as the result of a Web search.[1] You don't end up at the Wikipedia home page, you end up in an individual Wikipedia article. As UX practitioner Mike Atherton says (in conversation with Sara Wachter-Boettcher):

> Where once we built ourselves silos on the Web, these days it pays to recognize that it's really one Web and we're in the business of stitching our content into that wider canvas.
>
> —Sara Wachter-Boettcher, *Content Everywhere*[26]

Even major Q&A sites like Stack Overflow are distributed through search space. Of course, to post a question you have to go directly to the site. But you generally begin by searching for an answer. If you search on programming topics, Stack Overflow will come up often, reminding you to go there to post your question when search does not find an answer. But there is no compelling reason to search just Stack Overflow when there are so many other technical resources available on the Web.

According to David Weinberger, "Filters no longer filter out. They filter forward, bringing their results to the front."[27, p. 11] Few things are experienced at a fixed location on the Web. Most things are experienced in the ad hoc groupings caused by the Web's filtering forward of content related to the subject of your present interest.

---

[1] Wikipedia is large enough and comprehensive enough that some people do search it directly. However, most sites, probably including yours, are not that large or comprehensive.

This dynamic semantic clustering – bringing content together based not on its location but on its relevance to a particular search term or interest – is a key operational feature of the Web.

Dynamic semantic clustering is the key operational feature of the Web.

Even if your pages are behind a login or in a help system, they still appear as single pages to your users. Your pages are behind the login; users aren't. They are still operating in the context of the Web. They may have to do a separate search in your content set, but they are searching the Web as well and your page is just a page to them.

CHAPTER 4

Information Architecture Top Down

There are essentially two ways you can organize anything: top down and bottom up. In the book world, we organized everything top down, and for the most part we have brought that model to the Web. But does the top-down model work on the Web?

Book navigation

Traditional book organization is either linear or hierarchical. The organization is expressed through a table of contents (TOC), which is either a simple ordered list or a nested/hierarchical list like Figure 4.1:

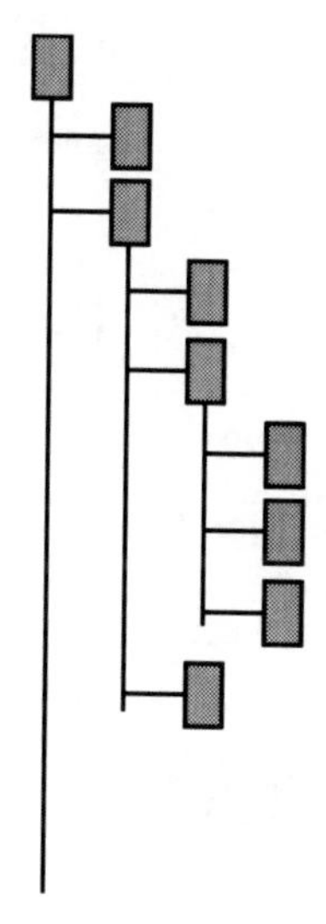

Figure 4.1 – Hierarchical TOC

In a top-down organization, you can see the whole organization in one piece, as though you were looking down on it from the outside.

However, the organization of a book is not apparent from an individual page. A page might contain headings that locate the page in a particular chapter or section, but it does not show you the relationship between that chapter or section and the rest of the book. For that you have to turn to the table of contents – a separate page or set of pages that describes the organization. The organization is visible from the outside, but invisible from the inside.

In a book, the organization is visible from the outside, but invisible from the inside.

Tri-pane help systems, like the one in Figure 4.2, change this somewhat. They place the TOC in a pane alongside the page. You can see where the page fits in the organization of the book because the content page and the TOC are side by side.

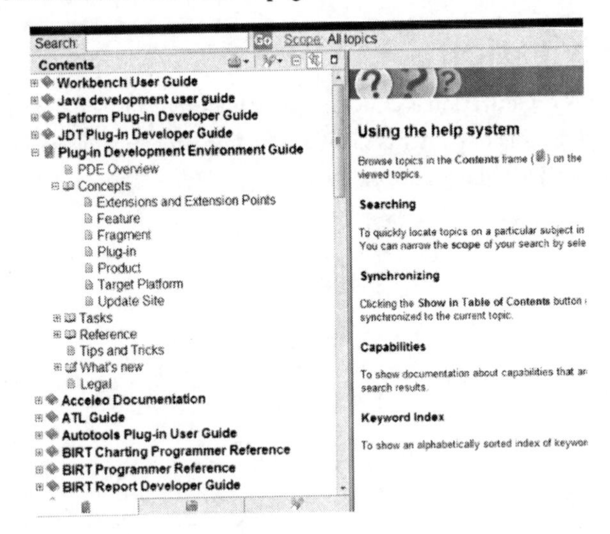

Figure 4.2 – A help system TOC

The trouble with TOCs

Looking at the table of contents in Figure 4.2, you can see it is not the TOC for a single book. Instead, it lists a library of books that includes multiple guides and references. When you move multiple volumes to the Web or set up a single comprehensive help system using the tri-pane help paradigm, you need to decide whether to put all the material into one TOC or create a separate TOC for each book (assuming you still think of your Web content as a collection of books).

Tom Johnson's blog post "Two Competing Help Models: One-Stop Shopping or Specialized Stores?"[1] explores two models for putting help information on the Web using the tri-pane help paradigm, and asks the question: Do you combine all the help into one site with a master TOC, or do you put up multiple small sites with separate TOCs? Multiple help systems make it harder to search or browse across the boundaries between systems, but:

> Browsing also becomes problematic when you have 4,000 topics in the table of contents (the one-stop-shopping model). Browsing through books and sub-books and sub-sub books and sub-sub-sub books to find the right topic is tedious.
>
> If you expand out all levels of the table of contents, the information starts to look really complex. Users may feel intimidated and overwhelmed about where to even begin. It's too much to learn. Subsequent forays into the help may lead to more familiarity, but not for new users.
> —"Two Competing Help Models: One-Stop Shopping or Specialized Stores?"

So, a large TOC is a problem when a reader approaches the content from the top down. Further problems become evident when we consider what is likely the more common case: a reader searches the Web for a specific answer and lands on a page in the middle of the help and *not* at the top of the help website. In classic tri-pane help, this will also land the reader in a particular place in the TOC in the left pane.

In Figure 4.3, the TOC shows where the page is located in the hierarchy of the help system, but it does little to locate the topic in the context of its subject matter. In folded form, this TOC is a high-level list of other subjects in the collection, few of which are likely to be relevant to the user's current task. If readers have questions arising from the people, places, objects, actions, and ideas mentioned in the topic, this TOC does little to help them find more information on those subjects. For the most part, pursuing those subjects in the content set means starting again at the top level.

---

[1] http://idratherbewriting.com/2013/02/13/two-competing-help-models-one-stop-shopping-or-specialized-stores/

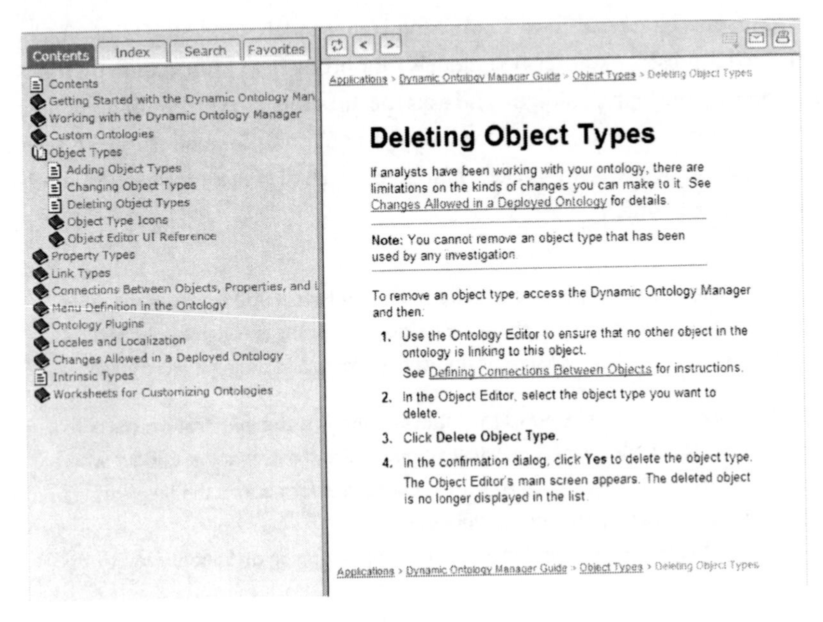

Figure 4.3 – Deleting Object Types example

Curriculum versus classification

So what is the function of a TOC? In some cases, the TOC describes a curriculum. In this case, the book is designed to be read sequentially, and the TOC outlines the order the author has created for the reader. A TOC can also be a manifest of independent objects that the user can choose to read in any order. Such a TOC can include either hierarchical topics or Every Page is Page One topics. A simple example of the latter is found in the *Popular Mechanics Complete Car Care Manual*, which is (most likely) a compendium of articles previously published in the magazine. It has a top level table of contents that groups articles by the area of the car (Figure 4.4).

Contents

Maintenance Basics 2
Engine 88
Drivetrain 138
Electrical and Electronic Systems 158
Chassis 214
Interior and Exterior 254
Appendix 314

Figure 4.4 – A TOC as manifest

Each of these area-of-the-car chapters has its own TOC (Figure 4.5).

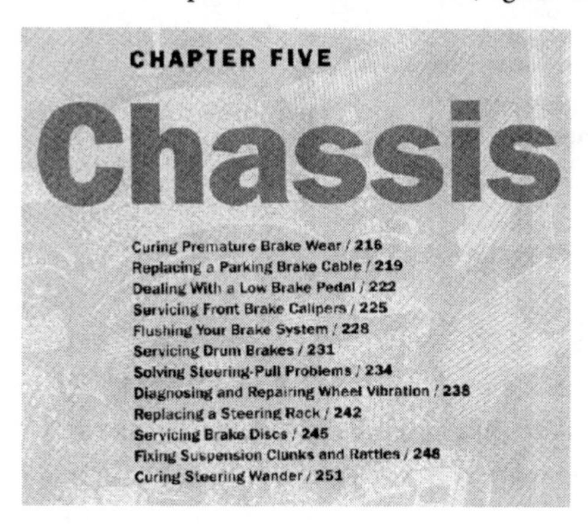

CHAPTER FIVE

Chassis

Curing Premature Brake Wear / 216
Replacing a Parking Brake Cable / 219
Dealing With a Low Brake Pedal / 222
Servicing Front Brake Calipers / 225
Flushing Your Brake System / 228
Servicing Drum Brakes / 231
Solving Steering-Pull Problems / 234
Diagnosing and Repairing Wheel Vibration / 238
Replacing a Steering Rack / 242
Servicing Brake Discs / 245
Fixing Suspension Clunks and Rattles / 248
Curing Steering Wander / 251

Figure 4.5 – Chapter-level of manifest TOC

Clearly, readers are unlikely to read straight through the *Popular Mechanics Complete Car Care Manual*. They will go directly to the article that covers the task they want to do, and the TOC helps them do that.

The WebMD symptom checker (Figure 4.6) is another example of a TOC that supports random access to content. It may not look like a regular TOC, but that is its function. Readers just point to where it hurts. Just as readers of the *Popular Mechanics Complete Car Care Manual* use the TOC to specify the area of their vehicle that is broken, users of WebMD use the symptom checker to specify the area of their body that is hurting.

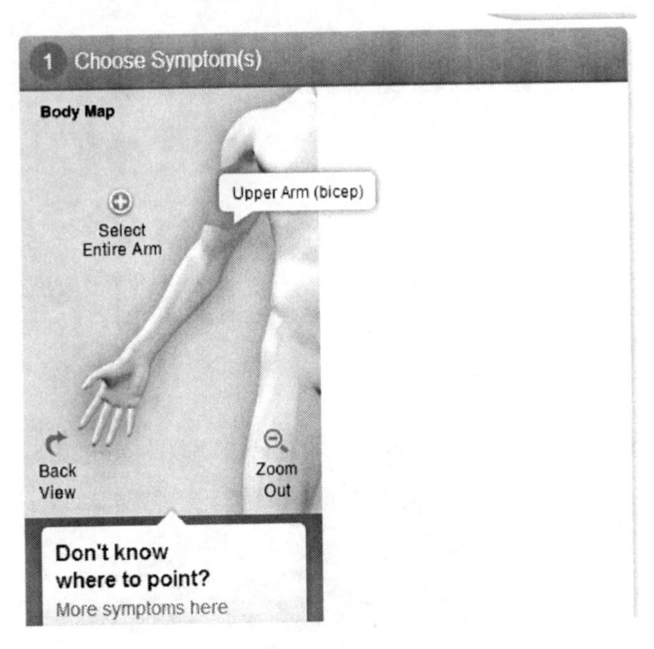

Figure 4.6 – WebMD symptom checker

Like the car care manual, WebMD is not designed to be read straight through or to facilitate the systematic study of medicine; it isn't WebMedSchool. Instead, WebMD helps people with particular symptoms figure out what's wrong. It is set up to be used in a focused, task-driven way by someone with a particular and personal goal.

While a random-access TOC could be an unordered list of articles, an unordered list can't get very big before it becomes unusable. Making the list alphabetical helps if the user knows the exact term to look for, but otherwise alphabetical is not functionally different from unordered. In such cases, a better way to proceed is to divide the content into groups, which is what we see in both the car care guide and the symptom checker. The car care guide groups individual maintenance items by parts of the car and the symptom checker by parts of the body. Both use a hierarchical organization.

Where the subject matter itself is naturally hierarchical, hierarchical organization works well. The home screen of the used-car site AutoCatch.com (Figure 4.7) presents users with a hierarchical selection of locations.

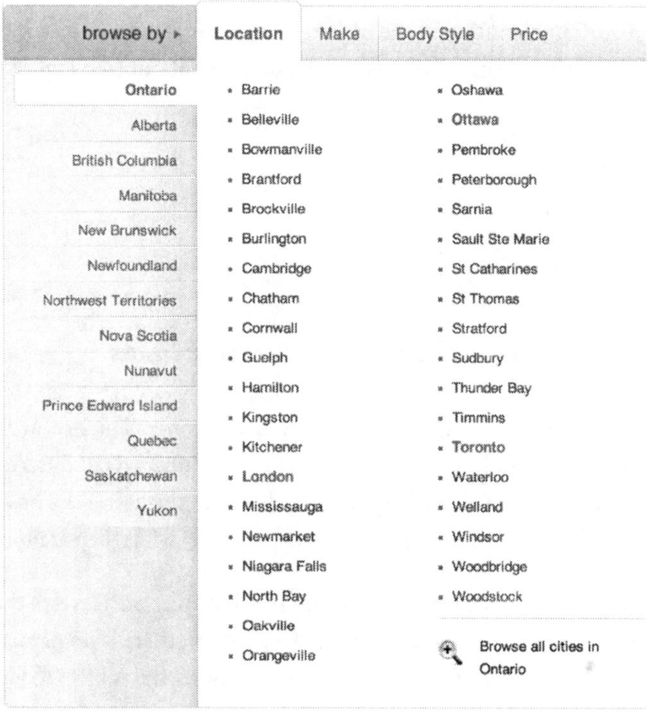

Figure 4.7 – AutoCatch home page

Where the subject matter is naturally hierarchical, hierarchical organization works well.

This works for two reasons: locations have a natural hierarchy (Ottawa is physically inside of Ontario), and used car buyers have a natural interest in the location of used cars. There are towns named Windsor and Woodstock in more than one Canadian province, but no one wants to look for used cars only in Windsor, Ontario and Windsor, Nova Scotia. Despite their common name, separating them in the hierarchy does not complicate matters for the shopper. The hierarchy works because it is natural, familiar, and appropriate to the task.

The limits of hierarchies

But what if the subject matter is not naturally hierarchical? What if the thing we are looking for is not naturally inside something larger the way Ottawa is inside Ontario? Imagine if AutoCatch continued the hierarchical organization of its content down through all the various aspects of cars that buyers care about (see Figure 4.8).

```
year
   transmission
      price
         mileage
            body style
               exterior color
                  private seller or a dealer
```

Figure 4.8 – An arbitrary hierarchy of used car characteristics

With that organization, a buyer who wanted to see all convertibles regardless of year, transmission, price, or mileage might have to look in hundreds of different places to see all the available convertibles. Play with the levels of the hierarchy however you like, buyers will still have to look in many places to find what they are looking for.

Here is where hierarchical organization breaks down. While each level of the hierarchy in Figure 4.8 is based on some useful aspect of a used car, there is no natural hierarchical relationship between levels. Rather than representing the subject it describes, this hierarchy distorts its subject.

The cultural bias toward hierarchies

We are culturally imprinted to organize using hierarchies. We use hierarchies to organize all kinds of things that are not hierarchical by nature. Figure 4.9 illustrates this with an example from a fun little book by Peter MacInnis called *The Speed of Nearly Everything: From Tobogganing Penguins to Spinning Neutron Stars*[17].

We are culturally imprinted to organize in hierarchies.

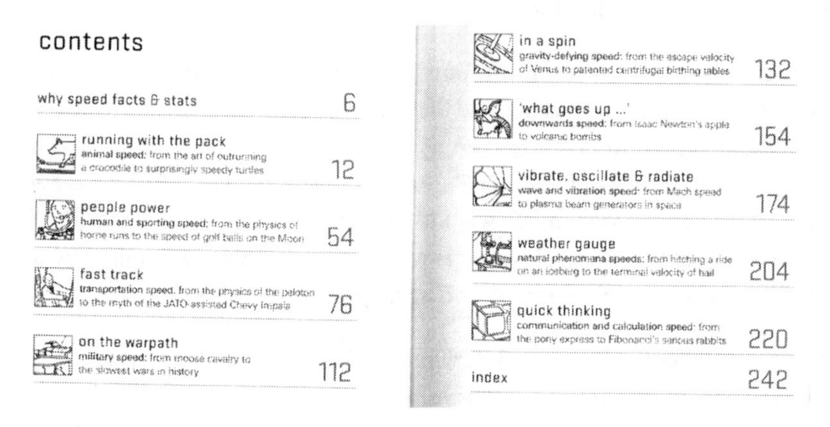

contents

why speed facts & stats 6

running with the pack
animal speed: from the art of outrunning a crocodile to surprisingly speedy turtles 12

people power
human and sporting speed: from the physics of home runs to the speed of golf balls on the Moon 54

fast track
transportation speed. from the physics of the peloton to the myth of the JATO-assisted Chevy Impala 76

on the warpath
military speed: from moose cavalry to the slowest wars in history 112

in a spin
gravity-defying speed: from the escape velocity of Venus to patented centrifugal birthing tables 132

'what goes up ...'
downwards speed: from Isaac Newton's apple to volcanic bombs 154

vibrate, oscillate & radiate
wave and vibration speed: from Mach speed to plasma beam generators in space 174

weather gauge
natural phenomena speeds: from hitching a ride on an iceberg to the terminal velocity of hail 204

quick thinking
communication and calculation speed: from the pony express to Fibonacci's serious rabbits 220

index 242

Figure 4.9 – TOC of The Speed of Nearly Everything

This TOC sorts entries into rough groups. But immediately we notice things that don't fit. If *running with the pack* covers animal speed, why is *moose cavalry* under *on the warpath*? And where would you look for the speed of a man on horseback: *running with the pack, people power, fast track,* or (thinking of cavalry) *on the warpath*? Why is the speed of golf balls on the moon covered under *people power* when *in a spin* covers gravity defying speed? And what about *what goes up ...* ?

Many of the hierarchies we use are unnatural and distort the subject matter. Why do we continue to use them then? The chief reasons, I suspect, are two fold:

Many of the hierarchies we use distort the subject matter.

First, our chief tool for organizing information for the last several centuries has been paper. You can express a hierarchy statically on a piece of paper, but paper is not good at expressing how subjects are related on multiple independent axes. Thus, we have adopted an organizational schema that our tools are capable of capturing.

Second, we tend to use analogs from the physical world to explain the digital world. The problem is, the physical world has constraints that don't exist in the digital world.

For example, consider two recent blog posts, "Structured Content is Like Your Closet"[3] by Val Swisher and "Content Strategy Can Save Us All From Slobdom"[4] by Meghan Casey, both of which use the analogy of a well-organized closet to illustrate how content management works today.

The basic principle of closet organization is to place like with like. Thus, we place shoes with shoes, coats with coats, and hats with hats. But there are problems with like-with-like organization. Things can be like each other in different ways, all of which matter at one time or another. For instance, you could have red shoes, red coats, and red hats, and you might want to wear these things together to create a co-ordinated red outfit. But in a closet organized by garment type, rather than by color, you have to go to multiple places to find a red garment of each type.

Other complexities arise: do you place winter and summer clothing together by type, or do you segregate the seasons and have two places for hats, two places for shoes, and two places for coats? In the physical world, there is no way out of this dilemma. When you store physical things together, you must choose one aspect as the principle axis of organization at the expense of other aspects. You can choose garment type, color, or season, but you can't choose them all equally. Whichever one you choose, you will artificially separate things that are like each other in one way because they are like something else in a different way. You have to sunder your red shoes from your red coat to place them with your other shoes.

This limitation does not exist in the digital world. Like can be stored with like for every aspect of alikeness without limit. Red socks can be stored with black socks on the basis of their sockness, with red gloves on the basis of their redness, and with winter wear on the basis of their warmth.

---

[3] http://www.contentrules.com/blog/structured-content-is-like-your-closet/
[4] http://blog.braintraffic.com/2011/08/content-strategy-can-save-us-all-from-slobdom/

In the physical world, putting something next to one thing means moving it further from another thing. In the digital world, you can put something near any number of other things in any number of dimensions. (In a few pages, we'll see how the site AutoCatch.com lets you group cars by any aspect of alikeness that matters to you.) Our experience of organizing things in the real world has conditioned us to think that the compromises the physical world imposes are inevitable and even natural.

In the digital world, you can put something near any number of other things in any number of dimensions.

The rise of the Frankenbooks

Neither of the TOCs in Figure 4.2 and Figure 4.3 defines a curriculum or a classification. Some of the items are references (classification) and some are guides (curriculum). Overall, both TOCs are hierarchical, but neither reflects any natural hierarchy in the subject matter, which probably doesn't have a natural hierarchy anyway. Once we start opening up lower levels in these TOCs, we are likely to find more of the same, with little hope of discovering what we need, except by accident.

And yet this is what we find in countless technical documentation sets across the web and, for that matter, in captive help systems. I call these creations *Frankenbooks.*

Frankenbooks seem to be on the rise. A single book made into a help system is bad enough, but now we are seeing help systems that comprise many books or many books worth of *building-block topics* threaded together. This leads to all-encompassing, deeply-nested information hierarchies in which all the pieces extracted from an original set of books have been threaded together into one monstrous info-glob.

A Frankenbook is organized neither for linear reading nor for random access. No matter where you land, you end up in a maze with buttons to move up, down, or sideways but no means of finding the end of any thread of exposition. Every page is page 297, and no page answers your question or helps you find a page that does. In a Frankenbook, the TOC represents neither a classification system nor a curriculum.

In a Frankenbook, the TOC represents neither a classification system nor a curriculum.

Frankenbooks have been with us for a while. For example, combining several unstructured books into a help system using an automated converter often produces a Frankenbook. But the growth of DITA seems to have created an increase in

Frankenbooks. Instead of Frankenbooks being created as a consequence of bursting multiple books, they are actually planned and assembled out of building block topics.

I am not saying that DITA adoption automatically leads to the creation of Frankenbooks. You can use DITA to do Every Page is Page One topic-based writing, to create help systems, or to create conventional books; you don't have to create Frankenbooks. The choice is yours. Nevertheless, for a tech pubs group transitioning to DITA with tight deadlines, limited funds, and a load of legacy content, creating Frankenbooks is often the path of least resistance.

The power of defaults is also at work here. Studies have shown that most people stick with defaults. And by default, the DITA Open Toolkit puts each topic in a separate HTML page. It is hard to imagine how it could default to anything else, since it can't guess at more appropriate groupings. However, if people create content from many small building-block topics, the resulting output will contain many small HTML pages, and many of those pages will not work well as page one. Defaults have the power to shape industries. As Bryce Roberts explains, Flickr's decision to default sharing to public rather than private changed the industry:

> This default to "public" had such a powerful effect on unlocking the network effects of this new service, and era, that it became canonized as a foundational principle of Web 2.0.
>
> —The Power of Defaults[6]

While we can't blame DITA for this default – and DITA provides several ways to change it – we shouldn't allow this default to lure us into making Frankenbooks.

Hierarchies simply do not scale.

Here's the hard truth: hierarchies become less logical and more arbitrary the larger they become. Look at any large TOC and arbitrariness, not logic, is the prevailing feature. Hierarchies simply do not scale.

[6] http://bryce.vc/post/28508177842/the-power-of-defaults

Faceted navigation

When a one-dimensional hierarchy is inadequate to organize your content, a faceted navigation may provide an alternative.

In faceted navigation you choose a value from column A and a value from column B, and the system shows you the items that match both selections. You can then select something from column C to narrow the result even further. Faceted navigation systems work interactively. Rather than making all your selections at once, you can start with a simple search on one factor. The system then shows you a list of results and populates the other selection fields only with the values from items that match the first query. This avoids the dreaded advanced-search phenomena where your first search returns no results, and you have to guess which search terms you need to remove to get a result.

Faceted navigation breaks down a complex search into small steps. At each step, you learn more about what results are available. This is great way to find what you need in a large collection. For instance, Figure 4.10 shows an inside page of AutoCatch.com:

Figure 4.10 – Faceted classification of used cars

Car buyers can view cars based on whatever features they find important: year, transmission, price, mileage, body style, exterior color, or type of seller (private or dealer). The AutoCatch interface allows buyers to select any facet in any order, and it tells buyers how many cars will remain in the list after applying a facet. If a particular facet would yield no hits, the interface won't even display it.

This scheme works because car buyers use these features to classify cars they might be interested in. For example, if a buyer has already decided to look for a sedan with fewer than 50,000 miles, with an automatic transmission, and not older than a 2008 model, AutoCatch will display cars that match these criteria. AutoCatch uses hierarchy where hierarchy works (selecting a location), and faceted classification where faceted classification works (selecting a car).

Another example of faceted classification can be found in movie rental sites that let you select movies by genre, star, rating, awards, live vs. animation, etc.

A simpler form of faceted navigation is the grouping of search results on Amazon. When you do a search on Amazon, it may return products from many different product categories (facets), ranging from books to movies to music to lawn and garden. But if you click on a category, the results from the other categories will go away.

It is notable that AutoCatch has a more complex faceted navigation interface than Amazon. AutoCatch is a relatively small site with just one product: used cars. Amazon is a huge site with many thousands of products, but it only provides a hierarchical browse-by-department capability and relies principally on search and the ability to limit search results by department or by author.

Clearly, scale doesn't drive the number of facets or Amazon would have many more facets than AutoCatch. The difference, I believe, is that car shopping is naturally a multi-factor exercise. Car shoppers have a set of well-understood, independent criteria for selecting a car. Book shoppers generally have only one principle criteria for buying a book – the subject it covers (or, occasionally, other books by a favorite author). Show me all the books on programming in Python, pressing wild flowers, or swimming with sharks (or by John Steinbeck). In some cases you don't even want to narrow by department – if you a buying a gift for a child who is into pressing wild flowers, you may want to see books from the book department and kits from the craft department side by side.

Facets must be native to the objects classified and of direct interest and familiarity to the audience. You must discover facets where they exist; you can't invent them. Facets work for used cars, but they don't work well for the range of products Amazon offers.

Most technical communication subject areas don't lend themselves to faceted navigation. However, one area where it might profitably be used is reference material. An API (application programming interface) routine has several familiar and interesting facets, including the data type of its parameters and return value. Programmers often need to find a routine that takes an argument of a specific type and returns a result of another specific type. A faceted navigation could easily be created to allow programmers to query an API reference this way.[8]

The limits of classification

The problem with classification as a principle of organization is that it only works when readers are familiar with the classification scheme, and most of the time they aren't. People do not say, "I have an ache in my second upper bicuspid," they say, "I have a toothache."

Experts classify things in their areas of expertise in order to be precise. Often, they make up vocabulary because common usage of common words does not divide the world neatly into categories. Thus biologists classify life into: kingdom, phylum/division, class, order, family, genus, and species. The botanist says "Plantae/Angiosperms/Eudicots/Rosids/Malpighiales/Violaceae/Viola/Viola tricolor," but most people say "Pansy." The expert classifies, everyone else names.

---

[8] I know of no examples of this actually being done. Please let me know if you know of one.

Figure 4.11 – Plantae/Angiosperms/Eudicots/Rosids/Malpighiales/Violaceae/Viola/Viola tricolor, AKA Pansy[9]

As Wikipedia explains:

> Many of these names play on the whimsical nature of love, including "Three Faces under a Hood," "Flame Flower," "Jump Up and Kiss Me," "Flower of Jove," and "Pink of my John."
>
> In Scandinavia, Scotland, and German-speaking countries, the pansy (or its wild parent Viola tricolor) is or was known as the Stepmother (Flower). This name rose out of stories about a selfish stepmother; the tale was told to children in various versions while the teller plucked off corresponding parts of the blossom to fit the plot.
>
> In Italy the pansy is known as *flammola* (little flame), and in Hungary it is known as *árvácska* (small orphan). In Israel, the pansy is known as the "Amnon v'Tamar." or Amnon and Tamar, after the biblical characters (II Samuel 13). In New York, pansies have been colloquially referred to as "football flowers" for reasons unknown. In some countries of Spanish language, the pansy is known as "Pensamiento" or "Trinitaria."
>
> —Wikipedia entry for Pansy[10]

[9] Image from Wikipedia Commons. Copyright © Aftabbanoori (CC BY-SA 3.0).

[10] http://en.wikipedia.org/wiki/Pansy

Name variation is one reason people use search. If you Google "stepmother flower" you will get a bunch of results about pansies. This isn't foolproof, of course. If you Google "football flower" you will get a bunch of results related to floral arrangements shaped like soccer balls or American footballs. Still, if you try to find "stepmother flower" by navigating the botanical taxonomy, you will get no results at all.

It's unlikely that a book on flowers will have all the common names for every flower in its index unless it is specifically about the variety of common names. But the Web has heard all of them.

The hard limit for navigation based on classification is this: people rarely classify their experiences or their questions. They simply name them or describe them. If you can classify your question, you can navigate to an answer. If all you can do is name it or describe it, you search.

People do not choose search because it's more discriminating or accurate than faceted navigation (it's not). It is not a question of which works better. They search because they can't classify their query; they can only name it or describe it.

If you able to classify your question, you can navigate to an answer. If all you can do is name it or describe it, you search.

Where top-down works

When does top-down navigation work and when doesn't it?

- It works when users want a curriculum. In this case they may well want an ebook or a paper book rather than a website, but either way, once readers have entrusted themselves to your guidance as author, a top-down curriculum-based TOC works.
- It works when you can provide a classification scheme that makes intuitive sense to the reader, ideally one where the reader has already classified the query before starting. This means the classification is natural to the subject matter in a way that matters to the reader. An imposed or artificial scheme won't work.
- It works when the classification scheme, natural or not, has been made canonical by a discipline or trade, has become part of the language of practitioners, and the audience consists entirely of fellow practitioners.

In fact, not only does top-down work in these cases, it may be mandatory. In these cases, if you don't provide top-down navigation, your users will probably be unhappy. If your users bring a classification schema with them, whether hierarchical or multi-faceted, your content had better be organized the same way.

It is no coincidence that most of the good examples of faceted navigation come from sites that sell stuff. As noted above, two things are necessary to make faceted navigation work. First, the reader must be familiar with the facets. Second, the reader must care about the facets independently. Those conditions are frequently met when you are selling things. They are rarely met when you are providing information.

An unfamiliar classification scheme has a lousy information scent.

Just because systems based on user classification schemes have succeeded, don't assume you will have the same success if you use a classification scheme dreamed up by an author, information architect, or content strategist. You can't expect your readers to learn your classification scheme as a prerequisite to finding the information they are looking for. An unfamiliar classification scheme has a lousy information scent.

For the writer, the information architect, and the content strategist, the lure of top-down classification is strong. Each of these people, to one extent or another, has to be concerned with the overall information set and its completeness, and these concerns are impossible to address without imposing some form of curriculum or classification on the content. The writer, the information architect, and the content strategist have to create a classification scheme to get their own work done, and their scheme quickly becomes the easiest and most reliable way for them to navigate the content set.

No wonder, then, that they believe that what works so well for them must also work well for the reader. But it doesn't. The writer, the information architect, and the content strategist classify their content because they work with it all day and think about it constantly. But the hierarchy that is so lucid to them is entirely opaque to their readers. It helps the creator find content, but doesn't help the reader. Navigating an unfamiliar classification system is like navigating a call center voice menu tree. (Keep hitting 0 until you get to a human.)

Teaching readers the classification scheme and then asking them to navigate the classified content usually won't work. It may work for frequent visitors with specialized needs that can't be met elsewhere, but, as content strategists frequently note, if you have to tell readers how to navigate your website, you have already failed.

Some people suggest doing a card-sorting exercise to come up with a classification that makes sense to users, and while that can lead to a more natural classification scheme, how you would classify one object within a particular set of objects is not necessarily how you would classify that same object in your own experience and your own set of concerns. More to the point, you generally would not expend the mental energy to classify it at all. Naming it is just easier.

When all they want is a single item, people simply don't care how that item fits in a classification; they just want the item. They expect, however unreasonably, that the answer should be as simple and straightforward as the question. And the question, as far as they are concerned, is always simple because it concerns one entirely concrete and discrete piece of their own experience.

What they can do, successfully, with that concrete and discrete piece of experience is type it into a search engine. Along with all the reasons mentioned in Chapter 2, this is why people prefer to search. And that leads us to the real heart of the website navigation problem: people do not arrive at the top – the home page – they arrive at the bottom. Every Page is Page One.

CHAPTER 5

Information Architecture Bottom Up

The essential architectural feature of the Web is that every page is a peer of every other page. Any page can link to any other page directly. Any hierarchical organization you try to impose overlays those peer-to-peer relationships, but the underlying architecture still asserts itself. A page can link directly to any other page, and a user can navigate directly to any page. The reader, and the page author, can route around any hierarchical organization. These direct connections between pages are what make the Web a web. Webs organize themselves bottom-up.

Sites like Amazon and Wikipedia make little use of top-down navigation. Once you arrive at an Amazon or Wikipedia page, whether by Google search, site search, or following a link, you will find it stuffed full of links. Wikipedia articles give you links to related subjects. Amazon pages give you links to other books or products based on your preferences, the preferences of others, and shopping patterns associated with the products themselves. You don't move up and down in these sites, you move sideways, from page to page.

These page-based links don't attempt to cover the whole of Amazon or the whole of Wikipedia. Both sites are far too big to cover effectively, and most of what they contain is not of immediate interest to the person viewing the current page. Instead, they link to things related to the current situation of the reader, the current subject of interest, and immediately related subjects. In other words, the navigation that these pages provide is local – it is not about the whole site, it is about the current subject and its related subjects. Like the results of a Web search, it is a form of semantic clustering.

Because content has traditionally been organized hierarchically, writers are often not familiar with organizing content from the bottom up, and so the idea of organizing content as a web seems foreign, chaotic, and dangerous to them.

One key aspect of a top-down organization is that the structure is apparent from the top but not from the bottom. The organization of a book is apparent from looking at the table of contents but not from looking at an individual page.

The Web stands this on its head. If you try to take a top-down view of the Web and map the peer-to-peer links in a set of related web pages, all you will see is chaos. Figure 5.1 shows what happens if you try to draw a map of a few pages:

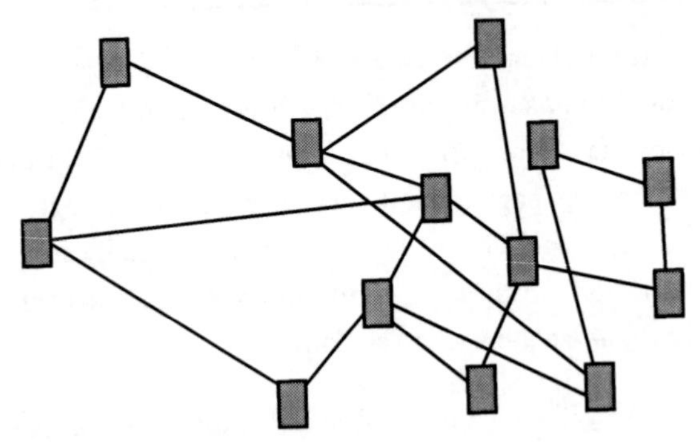

Figure 5.1 – A map of a simple web

With only a few nodes, the map remains comprehensible, but when you add more nodes and more links, all semblance of order and all hope of comprehension disappear from the top down view (Figure 5.2):

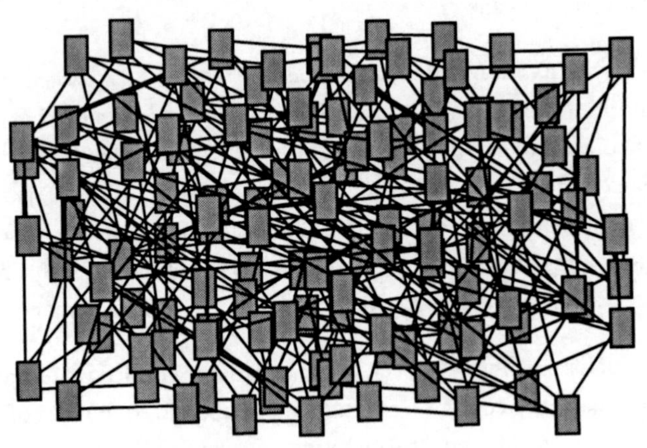

Figure 5.2 – Map of a complex web

For an author who has made a career writing books, this lack of a top-down view can be the most difficult thing to adapt to when starting to write for the Web.

However, in a properly constructed web of content, the organization is contained in the pages themselves. That is, by linking to each other and identifying themselves to the Web's filters with metadata, pages organize themselves into the Web. A page's position is cemented by links to it from other pages and selection by filters.

Web organization is always local, not in the sense that a page only links to pages in the same website but in the sense that the navigational and organizational tools are particular to an individual page and its subject matter. Each page has its own set of associations and affinities. Some of those affinities are local and some are distant, but they are all related to the current page. The page locates itself in a semantic cluster formed by links and keywords.

Paper encyclopedias have to be ordered alphabetically or topically, but ask how Wikipedia is ordered, and the question proves to be absurd. Wikipedia isn't ordered either topically or alphabetically. It is not ordered at all. It is connected.

A web of subject affinities

Wikipedia isn't ordered alphabetically or topically. It is not ordered at all. It is connected.

The Web's filters group content based on subject, scattering your content across the search space of the Web (as discussed in Chapter 3). Links connect content based on the affinities between subjects, connecting a topic on one subject to topics on related subjects. Thus filters and links work together to organize the Web.

Subject affinities are a feature of all writing, regardless of media. They anchor a piece of content in time, culture, and the reader's experience. Without subject affinities, content would float meaninglessly like a piece of nonsense verse: grammatically scannable, but semantically vacuous:

> 'Twas brillig, and the slithy toves
> Did gyre and gimble in the wabe;
> All mimsy were the borogoves,
> And the mome raths outgrabe.
>
> —Lewis Carroll, *Jabberwocky*

Your understanding of a topic depends on your knowledge of the subjects that the subject affinities point to – that is, the real world things that the text refers to. If all the words are just noise to you, you can't understand the topic in any meaningful way.[1] No subject, and therefore no topic, exists in isolation. Every subject is embedded in an environment, and your success in understanding and benefiting from the topic depends on your knowledge and experience of the environment in which its subject is embedded. But since many readers will not have a perfect grasp of that environment, they may need help acclimatizing themselves to it.

Traditionally, books have done almost nothing about subject affinities. The subjects are mentioned as they arise, but no connection is made. There would be very little benefit in providing such connections, because they would be too hard for the reader to navigate. Writers generally assumed (if they even considered the possibility) that readers would look these things up for themselves if they needed to. Helping them to do so was outside the writer's responsibility.

Links help keep readers inside your content set, as opposed to letting them wander off across the Web.

On the other hand, the Web makes it easy for readers to grab other resources to fill in gaps in their knowledge or experience. They don't even need links. If readers come across subjects they are not familiar with, they can quickly generate their own links using a search engine. However, links can help enormously. Following a link is less disruptive than doing a search and creates less cognitive overhead. As long as the link is to a useful resource, it makes life easier for your readers. Links also help keep readers inside your content set, as opposed to letting them wander off across the Web (which they are free to do even if your content isn't on the Web).

Consider the broad set of subject affinities in this excerpt from a Wikipedia article on the Manicouagan crater in Quebec (Figure 5.3):

[1] Yes, Humpty Dumpty did provide a glossary of some of the nonsense words in Jabberwocky, but that further demonstrates the point. If you don't recognize the subject affinities of a piece, you need additional material to make sense of them.

> The **Manicouagan Crater** is one of the oldest known *impact craters on Earth* and is located primarily in *Manicouagan Regional County Municipality* in the *Côte-Nord* region of *Quèbec, Canada,*[1] about 300 km (190 mi) north of the city of *Baie-Comeau.* At roughly 213-215 million years old, Manicouagan is one of the oldest large astroblemes still visible on the surface. Its northernmost part is located in *Caniapiscau Regional County Municipality.* It is thought to have been caused by the impact of a 5 km (3 mi) diameter *asteroid* about 215.5 million years ago (*Triassic* Period).[2] It was once thought to be associated with the end-Carnian *extinction event.*

Figure 5.3 – Introduction to the Manicouagan Crater article[2]

Here are just a few of the types of subject affinities in this article:

- Geographic coordinates
- Municipal location
- Place in time
- Relationship to historical events
- Scholarship
- Geology
- Astronomy

The first paragraph (Figure 5.3) sets the context for the article in relation to the real world (a key characteristic of Every Page is Page One (EPPO) topics that we will examine in detail in Chapter 10). The real-world reference points the article uses to set context include: physical location (Manicouagan Regional County Municipality, Caniapiscau Regional County Municipality, Côte-Nord, Quèbec, Canada, 300 km North of Baie-Comeau), age (213-215 million years, Triassic Period), classification and status (one of the oldest large astroblemes still visible on the surface), probable cause (asteroid), and historical association (end-Carnian extinction event). The topic has an affinity with all of these subjects.

---

[2] http://en.wikipedia.org/wiki/Manicouagan_crater Copyright © Wikipedia Foundation CC-BY-SA-3.0.

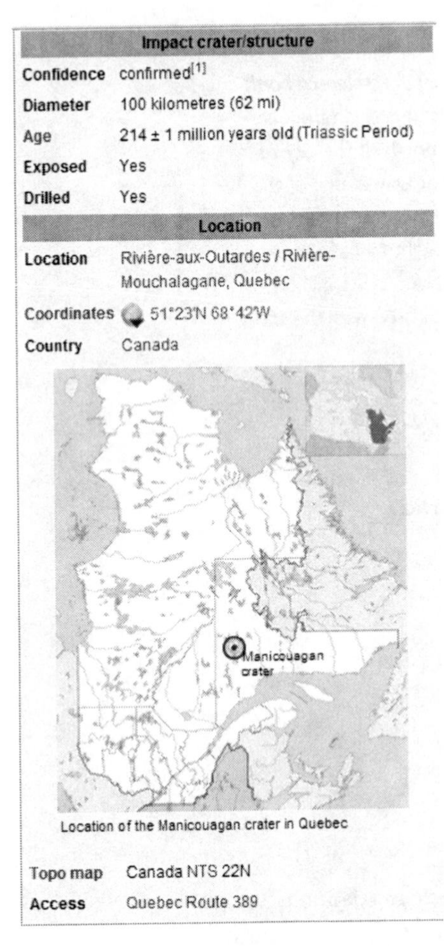

Figure 5.4 – Manicouagan Crater sidebar

Each of the subject affinities is furnished with a link (highlighted in *italics* in Figure 5.3), with the exception of "astroblemes" and "213-215 million years." The former is an omission, and the latter is remedied by a later reference to the Triassic Period. There are some anomalies in this passage (they can be attributed to the Wikipedia process and may have been rectified by the time you visit the page): the reference to the northern part being in the Caniapiscau Regional County Municipality looks like it was added later, and the two date references don't quite agree. Despite these minor anomalies, the contributing authors have created a good context-setting opening paragraph with strong links along the lines of subject affinity.

Figure 5.4 shows the right sidebar of the article. Most Wikipedia articles have a similar sidebar that contains reference information about the subject of the article. While these sidebars are not always identical across all articles of the same type (Wikipedia is not driven by a structured authoring system that would assure complete conformity), you will find they are broadly similar for articles on similar subjects. Similar subjects share properties and subject affinities, and topics about those subjects naturally conform to a type based on those shared properties (another common property of EPPO topics, and one we will examine in depth in Chapter 9).

We can see subject affinities expressed and linked in the sidebar, including: common characteristics of an impact crater, more location information (including exact coordinates), and links to an extensive collection of maps.

In addition to the standard subject affinities this topic shares with similar topics, it also contains information that is specific to the Manicouagan Crater. An example is shown in Figure 5.5, which discusses the theory that the Manicouagan Crater may be part of a multiple-impact event. The subject affinities here are: the other craters in the hypothetical multiple-impact event (all named and linked), the names and affiliations of the scholars who proposed the theory (linked where articles exist), and the mention of a similar multiple-impact event.

> It has been suggested that the Manicouagan crater may have been part of a hypothetical multiple impact event which also formed the *Rochechouart crater* in France, *Saint Martin crater* in *Manitoba, Obolon' crater* in *Ukraine,* and *Red Wing crater* in North Dakota. *Geophysicist* David Rowley of the *University of Chicago,* working with John Spray of the *University of New Brunswick* and Simon Kelley of the *Open University,* discovered that the five craters formed a chain, indicating the breakup and subsequent impact of an asteroid or comet,[4] similar to the well observed string of impacts of *Comet Shoemaker-Levy 9* on *Jupiter* in 1994.

Figure 5.5 – Manicouagan Crater article detail

V • T • E	Impact cratering on Earth	
Geology	Lists	Worldwide · Africa · Antarctica · Asia · Australia · Europe · North America · South America · by country
	Confirmed ≥20 km diameter	Acraman · Amelia Creek · Araguainha · Beaverhead · Boltysh · Carswell · Charlevoix · Chesapeake Bay · Chicxulub · Clearwater East & West · Gosses Bluff · Haughton · Kamensk · Kara · Karakul · Keurusselkä · Lappajärvi · Logancha · **Manicouagan** · Manson · Mistastin · Mjølnir · Montagnais · Morokweng · Nördlinger Ries · Obolon' · Popigai · Presqu'île · Puchezh-Katunki · Rochechouart · Saint Martin · Shoemaker · Siljan Ring · Slate Islands · Steen River · Strangways · Sudbury · Tookoonooka · Tunnunik · Vredefort · Woodleigh · Yarrabubba
	Topics	Alvarez hypothesis · Breccia · Coesite · Complex crater · Cryptoexplosion · Ejecta blanket · Impact crater · Impact structure · Impactite · Cretaceous–Paleogene boundary · Late Heavy Bombardment · lechatelierite · Meteorite · Moldavite · Ordovician meteor event · Planar deformation features · Shatter cone · Shock metamorphism · Shocked quartz · Stishovite · Suevite · Tektite
	Research	Baldwin, Ralph Belknap · Barringer, Daniel · Barringer Medal · Chao, Edward C T · Dietz, Robert S · Earth Impact Database · Hartmann, William K · Impact Field Studies Group · Lunar and Planetary Institute · Melosh, H Jay · Ryder, Graham · Schultz, Peter H · Shoemaker, Eugene · Traces of Catastrophe book
Astronomy	Observation	Asteroid · Catalina Sky Survey · Close approaches · Earth-crosser asteroid · Impact event · LINEAR · LONEOS · Meteoroid · Meteoritical Society · NASA NEAT · Near-Earth Object (NEO) · NEOSSat · NEOCam · Orbit@home · OSIRIS-REx · Palermo Scale · Pan-STARRS · Planetary science · Potentially hazardous object · Sentinel · Sentry · Spacewatch · Torino Scale · WISE
	Defense	Asteroid impact avoidance · B612 Foundation · Gravity tractor · Ion Beam Shepherd · Japan Spaceguard Association · NEOShield · Spaceguard · The Spaceguard Foundation
	Potential threats	1950 DA · 1994 WR_{12} · 1999 RQ_{36} aka 101955 Bennu · 2002 MN · 99942 Apophis · 2007 VK_{184} · 2009 FD · 2013 BP_{73}

Figure 5.6 – Manicouagan Crater article footer

Perhaps the most interesting example of setting a topic in its subject context – and of formally managing and linking along lines of subject affinity – is the article footer shown in Figure 5.6. This footer, similar to ones found in many other Wikipedia articles, places the Manicouagan Crater into the taxonomies of the fields of study that encompass it: Geology and Astronomy. Note that this article's place in these taxonomies is not managed externally in a site taxonomy, it is expressed locally in the article footer. And the footer is confined to those aspects of the taxonomy that are related to the subject affinities of the article itself. This turns the article into a hub of its immediate area in its subject space, the crossroads of a semantic cluster.

Irregular subject affinities

In many cases, some of a topic's subject affinities will be irregular. Irregular affinities are relationships that are not shared by other topics of the same type. That is, they belong to the instance, not the type. For example, the Manicouagan Crater topic has a subject affinity with the Open University because scholars there came up with an unusual theory about the crater's origins. Not every topic on an impact crater will have that affinity.

Irregular affinities abound. While most topic types have a set of affinities that occur in every topic of the same type, any individual topic can have irregular affinities. Just because these affinities are irregular does not mean their subjects are of secondary importance. Often the most critical pieces of content we create are the ones that deal with special and particular relationships between subjects.

Regular affinities tend to be intuitive and easy to discover and anticipate. When someone gets stuck, it is often because the topic has some peculiar and irregular relationship with another subject. The irregular affinities, then, are often the most critical ones, the ones that are most important to findability and comprehension.

Indeed, this is one of the great beauties of organization and linking based on subject affinities. Readers should neither know nor care whether a subject affinity is regular or irregular, especially if it is irregular with respect to some arbitrary classification scheme. In bottom-up, link-based navigation there is no distinction between regular and irregular subject affinities or in the links that express those affinities. A classification scheme may well have been at work behind the scenes organizing the content and governing its types and vocabulary, but all of this is rightly hidden from readers, who just want to know a little more about X.

Subject affinities are not citations

While almost all of the subject affinities we found in the Manicouagan Crater article are linked to other Wikipedia articles or closely related resources such as GeoHack, the article does not talk about these other articles as articles. These links are not links to referenced content such as you would find in a footnote or bibliographic reference, they are links to referenced subjects, which happen to lead to Wikipedia articles, but which could just as well lead to any other resource.

Subject-affinity links neither recommend the content they link to nor cite that content for justification, they simply give an interested reader a place to go for more information. The original articles don't depend on any specific content or argument contained in the articles they link to. As long as the linked-to articles continue to be relevant to their original subjects, edits to them will not affect the validity or function of the links.

Therefore, subject-affinity links are not citations. Wikipedia does use citations, of course. In fact, it tries to make sure that every assertion in every article cites a source to back it up. But it does not use links for citations. It uses footnotes. Often the same word or phrase in a Wikipedia article will have both a subject-affinity link and a citation footnote. Because of Wikipedia's breadth, most subject-affinity linking stays within Wikipedia, and lots of other sites use it for for this purpose as well. This book also uses Wikipedia for this purpose.

Every topic in your content set shares multiple subject affinities with other topics in your content set and across the Web. Readers who navigate through your content do so to get more information on subjects of interest to them, and they will follow subject-affinity links to pursue those interests. Linking within your content can encourage readers to stay within your content. If you don't provide links or you don't provide content on a subject of interest, your readers will search for it themselves and leave your content. You can't keep readers by declining to link, but you might keep them with good content and subject-affinity links to relevant material.

Topics as hubs

Every topic is a hub of subject affinities that place it in a subject space. A well-constructed Web topic shows how the Web is organized around it and allows navigation along multiple lines of subject affinity.

Therefore, creating a collection of Every Page is Page One topics is not like publishing a book. Your topics will be scattered across the Web's search and social spaces, which is what you want because it gives you far more visibility than if you were confined to your one tiny corner of the Internet. But it also means that the world will not see your site as a single entity; it will see your pages. Each page is the hub from which all subsequent exploration of the content will begin, and upon which your ability to influence readers depends.

Your off-Web content lives behind a one-way door.

Even if your content is not on the Internet, that just means fewer people will find your pages in their search results. Those who do have access to your content will still see individual pages and will still read them in the context of the Web. Your pages are still hubs just as surely as if they were on the Internet. In fact, it is even more important that your pages be effective hubs within your own content set, because off-Web content is behind a one-way door. Once readers leave your content for the Web, the only path back is through your portal (login, subscription, or whatever), which means there will be few, if any, direct links for readers to follow from the Web back to your content.

The flattening problem

Any two-dimensional presentation of organizational relationships (on paper or on a screen) will, by necessity, flatten them. After all, the medium is flat. We are also limited in our ability to visualize and express complex multi-dimensional relationships. Whether that limitation is innate or the result of thought patterns developed by our five-thousand-year experience with paper, we flatten as we attempt to understand.

When we flatten, we distort. Things that should be close together are forced apart, things that should be distant are forced together, angles are twisted, three and four dimensions are compressed into two. Even if flattening is necessary for comprehension, it still distorts the thing we seek to comprehend. This distortion can greatly increase the cost of understanding and acting on information.

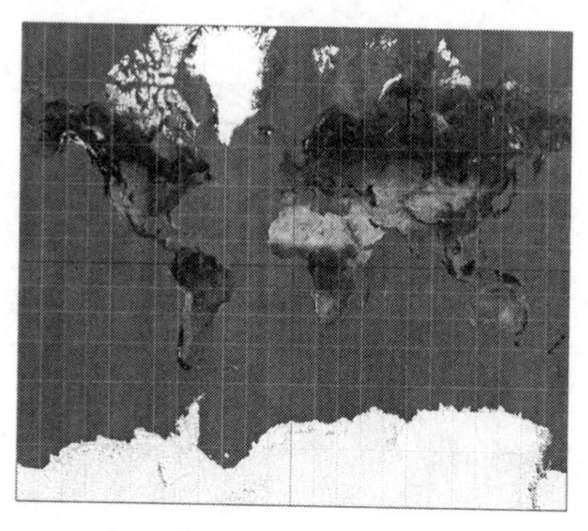

Figure 5.7 – Mercator projection[3]

One of the oldest examples of the problems with flattening can be seen in attempts to map a round planet on flat paper. Mapmakers have produced a variety of projections, each of which distorts the reality of the world in a different way. The classic example is the Mercator projection,[4] which exaggerates the size of Europe compared to Africa. It is also particularly generous to Canada and makes Alaska appear almost as large as the entire continental US. It represents the poles, which are points, as lines equal to the length of the equator, never lets the lines of longitude meet, and distorts the distances between the lines of latitude. Despite all these distortions, however, the Mercator projection is valuable for marine navigation because it makes rhumb lines[5] straight. In a flattened view of the world, we distort everything else to make a single property easy to work with.

While a globe represents the world accurately, it can be more difficult to use. Mercator and other projections may distort, but they present the world in a way that our brains find easier to deal with.

[3] Copyright © Wikipedia user Strebe, CC-BY-SA-3.0

[4] http://en.wikipedia.org/wiki/Mercator_projection

[5] http://en.wikipedia.org/wiki/Rhumb_lines

Figure 5.8 – Perspective Drawing[6]

Perspective drawing (Figure 5.8) deliberately distorts the size of distant objects compared to foreground objects to create the illusion of depth, but in doing so it destroys the ability to judge relative sizes. The eye accepts the illusion, but sometimes, for technical purposes, we need to flatten differently.

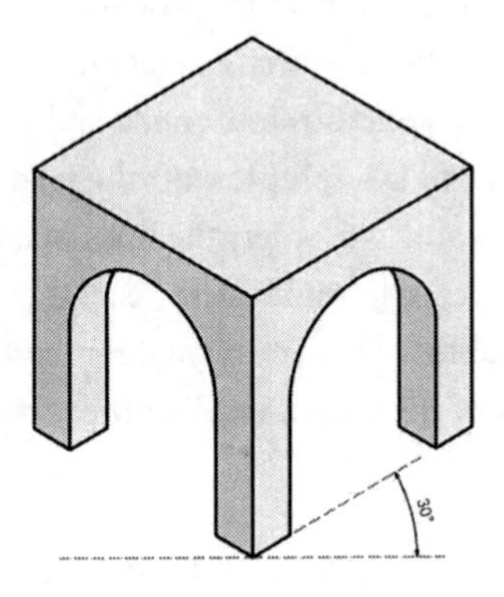

Figure 5.9 – Isomorphic Drawing[7]

[6] Image: Wikimedia Commons, Copyright © Wolfram Gothe, CC-BY-SA-3.0.

[7] Image: Wikimedia Commons, Copyright © Matteo Carcassi, CC-BY-SA-3.0.

For example, isometric projections (Figure 5.9) preserve dimensions but distort angles and perspective. Again, we flatten in different ways, distorting everything around a single axis that we find useful for a single purpose.

In text and online, we have built a civilization and a science on the flattening of information. This does not change the limits that flattening imposes on our powers of expression and on our ability to convey and acquire real understanding of the world. But these restrictions belong to an age in which the predominant tool for the extension and sharing of our intellect was paper.

We have a new tool now, one that is not restricted in its capacity to capture and express multiple dimensions of information. Computers and, perhaps more importantly, computer networks, are not dimensionally restricted in the way paper is. They allow us to represent and explore worlds and problems in multiple dimensions. We can represent a multidimensional world in the computer's memory and create algorithms to represent that world without distortion. We no longer have to flatten to represent. We no longer have to flatten to understand.

And we no longer have to flatten the organization of content to make it navigable. We can make it navigable in as many dimensions as necessary to represent all dimensions of subject affinity, including both regular and irregular subject affinities.

In *Big Data: A Revolution That Will Transform How We Live, Work, and Think*[19], Viktor Mayer-Schönberger and Kenneth Cukier make a similar point about tagging, a common means by which users can note the subject affinities of objects on the Web.

> The imprecision inherent in tagging is about accepting the natural messiness of the world. It is an antidote to more precise systems that try to impose a false sterility on the hurly burly of reality, pretending that everything under the sun fits into neat rows and columns. There are more things in heaven and earth than are dreamt of in that philosophy.[19][8]

[8] I am indebted to Tom Johnson for finding this quote, which he featured in his blog post When Organizing Big Data Content, It's Okay To Be Messy [http://idratherbewriting.com/2013/08/07/-big-data-when-its-okay-to-be-messy/]

In other words, the world does not fit our neat categorizations. And our technology no longer forces us to organize our content or our experience that way. If bottom-up organization appears messier than top-down organization, it is only because it more accurately reflects the messiness of the real world.

Of course, readers still read text as a linear sequence and see drawings projected on a flat screen. But we now have the ability to allow readers to rotate their view and, through linking, traverse the information space along multiple axes, as the Manicouagan Crater article provides many different directions for readers to follow up related subjects from geography to astronomy to geology to reservoirs to tourism. We don't have to give them flattened content. We can give them a projection of multidimensional content and the tools to manipulate the projection as they see fit.

Broader, deeper, more dynamic

This is a profound change from how we are used to organizing content. Traditionally, there has been one navigational schema for the entire content set, and each piece of content has been fitted into a pre-ordained place in that schema. Navigation is fixed and external. Content hangs off the hierarchical table of contents like ornaments hang off a Christmas tree.

In the Web model, each topic is a navigation hub in its own right, and topics organize themselves by expressing their subjects and linking to other topics along the lines of subject affinity. The navigational view from every node is specific to that node.

When an information set grows beyond a certain size, there is no other way to make it properly navigable. Seeing the entire set at once is too overwhelming, and artificially segmenting it is too confining. Therefore, navigation can't be universal, it must be specific to the subject locale and must change every time you change subjects. This way, you can have an arbitrarily large information set with manageable navigation at every point and still have the ability to travel wherever subject affinities lead you.

That is the broader point I am making here. Topics don't fit in one place in one grand classification. They exist at the intersection of multiple lines of subject affinity. The secret to successful web navigation is not to guide readers into a single classification scheme, but to allow them to travel at will along any of the lines of subject affinity that interest them.

Topics don't fit in one place in one grand classification. Instead, they exist at the intersection of multiple lines of affinity.

Top-down navigation fails because it is a static grouping on a system characterized by dynamic semantic clustering. To put it another way, if the Web is a collection of filters, the navigational schema of your website is just one more filter out of the many filters that may return (or reject) individual pages of your content. It is likely that visitors reach your pages (not your site) using some filter other than your site's top-down navigation. Consequently, your ability to influence, help, and retain visitors begins on that page, and starting from that page, their navigation will be bottom up.

This is the hurdle technical communicators must overcome to deliver useful, navigable content on the Web. We need to break away from the top-down organizational schema of the book world and learn to adopt, create, and manage a bottom-up organizational schema that supports an information collection as vast, vibrant, and fluid as the Web.

Should we abandon top-down navigation?

Does this emphasis on bottom-up navigation mean that you should abandon top-down navigation entirely? Certainly not. Many users will arrive at internal pages of your site via searches or links, and those pages should work as page one and act as a navigational hub. But some visitors will arrive at your home page and attempt to navigate your site. You need to accommodate these visitors, too.

In doing so, of course, you should recognize the limits on top-down navigation techniques that I described in Chapter 4 and try to create top-down navigation that is usable. The thing is not to prefer one navigational technique or preference exclusively over the other, but to recognize the diversity of audience needs and habits and accommodate both. One of the great things about bottom-up navigation is that, because it is distributed, and always local in the context of any regular page, it is also largely unobtrusive and should work well within any top-down framework.

Top-down navigation works better if it doesn't need to carry the whole navigational burden. Readers who arrive at the home page and navigate from there reach inside pages after the first click. If those inside pages provide good bottom-up navigation, that takes some of the burden off the top-down navigation. By allowing the two to work together, you can often simplify the top-down navigation, making it easier to understand. In the presence of good bottom-up navigation, top-down navigation does not have to be comprehensive to be effective. You can avoid the scary monster TOC and still provide effective access to all of your content.

Every Page is Page One is not about pitting one means of navigation against another. Instead this approach recognizes that any page a reader lands on is page one for that reader. You can't assume anything about how your reader arrived at that page – it could have been through search, a link, or your top-down navigation. You just need to write that page as if it were page one.

The role of lists

Bottom-up organization does not mean that there is no role for lists of resources. In fact, lists play a big role in bottom-up organization. A list is not the same as a table of contents, because it is not a manifest of a container. It does not list contents. It lists items of interest on a particular subject. (It is a semantic cluster.) Like all bottom-up navigational devices, lists are local and defined by subject.

In the section titled "Curriculum versus classification" we saw that most books attempt to prescribe a curriculum for readers, whereas an EPPO topic generally assumes readers will construct the curriculum. But that does not mean you can't offer a suggested curriculum using a list. The topics listed in such a curriculum may or may not have been written to be part of the curriculum, and the authors may or may not know that their topics are on the list. The list is simply an ordered curation of available material on a subject.

Your curriculum is just another web page – another topic in a web of topics. A curriculum topic is very similar to the pathfinder topic described in Chapter 15.

Lists are common on the Web and can be considered a type of Every Page is Page One topic. For example, Wikipedia has lists on a huge variety of subjects. It also uses lists within topics, particularly in footers, as we saw in Figure 5.6. Amazon lets a user compile lists of books on any topic. It also creates recommended lists of books based on that user's search terms and on what other people who bought the current book also bought or looked at. Curation sites and services are simply generators of lists. Even search engines really do nothing except generate lists of potentially relevant pages. Faceted navigation systems are essentially systems for narrowing down lists.

Lists are a type of Every Page is Page One topic.

A list is open ended. It does not have to own the things it points to, and it does not have to include anything that it not related to the subject of the list. Lists can, and often do, connect things that exist in multiple domains and that are created and owned by multiple people.

You don't have to create lists using a top-down approach. Lists are just nodes in the network. Lists can point to topics or other lists, just as topics can point to other topics or lists. Wikipedia even has a List of Lists of Lists.[9] Like any topic, a list is a hub of its local area of the network.

Lists can provide a connective layer between higher- and lower-level topics. For instance, consider the Wikipedia List of Stock Characters,[10] which exists, in a sense, between the general article on stock characters and articles on individual stock characters such as the Manic Pixie Dream Girl.[11] In a hypertext environment, such lists both collect instances of a type and provide bottom-up navigational support for exploring that type and its instances.

[9] https://en.wikipedia.org/wiki/List_of_lists_of_lists
[10] http://en.wikipedia.org/wiki/List_of_stock_characters
[11] http://en.wikipedia.org/wiki/Manic_Pixie_Dream_Girl

Characteristics of Every Page is Page One Topics

What does an Every Page is Page One topic look like? How do you write one? How do you know if you've written a good topic? This part covers the purpose and anatomy of Every Page is Page One Topics.

CHAPTER 6

What is a Topic?

We have seen that in the context of the Web, readers move around from one information source to another, often navigating more from the bottom up than from the top down, following the scent of information. When readers behave this way, every page becomes, for them, a new page one. The challenge for writers is to produce effective Every Page is Page One topics with a strong information scent. To understand what it means to write a good Every Page is Page One topic, we should start by clearing up some ambiguity about what the word topic itself means.

The word *topic* is used in many senses and in many fields, but for technical communicators *topic* refers to a small independent piece of information on a single subject. However, within that broad definition you will find significant variation in the scope and size of topics, degree of independence of topics, and the nature of relationships among topics. I coined the term *Every Page is Page One* topic (or EPPO topic) to distinguish my use of the term from other uses.

I don't claim that other definitions are incorrect; topics are designed for a variety of different, useful purposes. And I won't attempt to cover all possible definitions of topic; there are many and they overlap considerably. Instead, I will focus on two broad classes of topics used by technical communicators (the terminology is mine): building-block topics, and presentational topics

Building-block topics

Building-block topics are designed to be built up into other, larger information products. Andrew Brooke[1] gives a clear description of the building-block approach in his blog post "Topical Docs"[4] in which he compares topics to electrons:

[1] Andrew Brooke is a Senior Technical Writer in Toronto, Canada

> 1. A topic is to a document what a subatomic particle (such as an electron) is to matter. It is the basic component in a document. Each topic can and must stand alone.
>
> —Andrew Brooke, "Topical Docs"[4]

What does "stand alone" mean here? A brake caliper is a basic component of a car. Does the brake caliper stand alone? Certainly it can stand alone on the shelf at an auto parts store, but it serves no useful function until it is attached to the car. In this sense, it does not stand alone – it can only perform its function when integrated into a larger system.

Brooke goes on:

> 2. Combinations of topics are like atoms. They form a *section* of a document containing a group of related topics. This corresponds to a book within an online help TOC, or a chapter within a book.
>
> 3. Groups of sections are like groups of atoms, or *molecules*, for example, a water molecule. These correspond to an entire document.
>
> 4. Groups of documents form a library, which is like the various molecules combined together to form the complex matter, or *compounds*, that we encounter every day, everything from plastic to clothes to hamburgers.

A building-block topic must fit the context it is placed into seamlessly.

So, a building-block topic is a component of a book in the same way that a carbon atom is part of a hamburger or a brake caliper is part of a car. A building block has to exist separately, but it doesn't have to function separately. You cannot lunch on a carbon atom or go for a ride on a brake caliper. They do not stand alone functionally. They are building blocks, not finished units. Authors use building-block topics to assemble larger units of content. Readers should rarely see a building-block topic in isolation. Therefore, a building-block topic should seamlessly fit the context it is placed into.

There are two ways to approach this: context-dependent and context-free. Context-dependent building-block topics must be preceded by a lead-in topic and followed by a lead-out topic. When you select a topic for reuse, you need to place it next to topics that provide the correct lead in and lead out. You often get context-dependent topics if you mechanically chunk an existing book. These topics are like jigsaw-puzzle pieces – they only go together in a certain order.

The other approach is to create context-free building-block topics. You can use context-free building-block topics in many different places. You still need to provide some context, but you don't need to use a specific lead in and lead out. You can think of these topics as being like Lego blocks – you can put them together in any order, but only some orders will make sensible, usable constructs.

Presentational topics

A presentational topic is designed to be a unit of presentation. That is, it is intended to be the unit a reader receives. Clearly an Every Page is Page One topic is a presentational topic, but not all presentational topics are necessarily Every Page is Page One topics. Some presentational topics are meant to be read, or at least navigated, in a particular order or hierarchy, which they depend on for some or all of their context.

You will often find presentational topics of this sort in a help system, particularly one authored in a HAT (Help Authoring Tool). You could think of this class of presentational topics like playing cards – independent, but functioning as part of a deck and strongly related to other cards by number, rank, and suit. They may be dealt in different orders and combinations, but separated from the deck they lack some of the context that gives them meaning.

Every Page is Page One topics

Every Page is Page One topics (EPPO topics) are presentational topics that are meant to function alone, without dependence on a hierarchical structure. EPPO topics work equally well no matter how readers get to them. In his blog post "It's help, but not (quite) as we know it"[20], Scott Nesbitt praises Google's approach to the documentation for Chrome:

> One of the first things that I noticed was the way in which the documentation was described. *Help articles*. Yes, *articles* and not *documentation* or *user manual* or *online help*. That's a very subtle (or maybe not) distinction. But it's a distinction that can be psychologically powerful.

This changes what it means for a topic to stand alone. An article can be read on its own. It is not part of a larger manual. It stands alone not merely by existing separately, but by functioning separately.

Nesbitt goes on:

> Ask most people how they learn to use software or hardware, and I can bet that a majority say that they don't use the so-called *official* documentation. Most turn to search engines or sites like Lifehacker, eHow, or Make Tech Easier.
>
> By labeling the documentation as a set of articles, Google is (whether consciously or not) positioning the documentation for its products like articles that appear on the sites that I listed in the previous paragraph.

And, like most of the helpful material on the Web, these sites consist of articles, blog posts, forum posts, and so forth, all of which function independently.

An Every Page is Page One topic is designed to establish its own context and to function independently. This does not mean that Every Page is Page One topics cannot belong to collections, but those collections will tend to be organized bottom up rather than top down. You could think of a collection of Every Page is Page One topics like a box of toy cars. Each car is an independent toy in its own right. You can add a new car to the box or take an existing car away, but you still have a box of cars. And all the individual cars still work fine.

This does not mean that a topic has to be part of a bottom-up collection to be an Every Page is Page One topic. Precisely because it is independent enough to function as page one, a topic can remain an Every Page is Page One topic even if it is included in an hierarchical help system or treated as a building-block topic in a book. In fact, it is not unusual to find help systems today that are a mix of Every Page is Page One topics, hierarchically dependent presentational topics, and even building-block topics presented on their own.

Similarly, it is not a given that building-block topics are always used to build books. They may be used to build books or presentational topics, including Every Page is Page One topics, or to build brochures, catalogs, or email campaigns. The whole point of a building-block topic is that it can be used to build a variety of things. You don't have to choose between presentational topics and building-block topics. You just need to clearly understand which you are writing, or talking about, at any given time.

Economics and the evolution of topics

At one time, technical communicators wrote user guides as books and help systems as collections of presentational topics. Unfortunately, cost pressures made this approach unaffordable for many, and writers began using tools that created a help system by *bursting* the user guide at section boundaries. The topics created by this method were just arbitrarily separated sections of the book, and help systems were reduced to being just another navigation scheme for the user guide.

Under these circumstances, it was perhaps inevitable that the word *topic* would come to mean a chunk of a book (though such a topic is usually neither a good building-block topic nor a good presentational topic). In the continuing search for efficiency, people began to look at ways of sharing authoring responsibilities and reusing content across multiple information products. They came to see help topics as a more efficient unit to author and share. Rather than write books and burst them into topics, they would write topics and build them into books (and other information products).

DITA and Information Mapping

DITA[22] cemented this use of *topic* in the tech pubs lexicon. DITA was influenced by *Information Mapping*[15], borrowing the idea that a document is a map connecting different types of content objects. In Information Mapping, the content objects are called *blocks*. Blocks are not intended to stand alone, and there is no theory of reuse around them. They are simply the elements that make up a document. The purpose of mapping a document is to make the document easier to understand by writing each block succinctly and clearly and connecting blocks in the right sequence to ensure readability. Information Mapping is a design philosophy for creating documents using blocks and maps: the document is the intended output, not the map or the block.

DITA changed the terminology and the focus. Blocks became *topics*, and the six major block types of Information Mapping (concept, procedure, process, principle, fact, and structure) were reduced to three base topic types (concept, task, and reference). Maps became not a design philosophy but a mechanism for assembling books (or other information products) out of building-block topics, or, alternately for organizing a collection of presentational topics hierarchically.

DITA does not actually specify whether its topics types are intended to be building-blocks or presentational, and in practice you may find them used both ways – as building blocks of a book and as individual help topics in a help system.

Topics and the Web

While all this was going on in the technical communication world, the Web was becoming the largest collection of Every Page is Page One topics anywhere. These topics were not parts of books or help systems. They were nodes in a world-spanning hypertext network. They were connected and made searchable so that people could find and use them as needed. The Every Page is Page One topic is the natural and dominant form of topic on the Web.

Technical communication was one of the earliest and most vibrant forms of content on the Web. Before the development of the Web, Usenet was a vast collection of discussion groups that covered every imaginable technical subject and gave birth to the concept of the FAQ as we love and abuse it today.[2] This tradition of communal technical communication continues on the Web in countless forums, technical blogs, Q&A sites, and online magazines, all of which contain Every Page is Page One topics.

[2] In the Usenet days, a FAQ actually was a collection of the questions that had been most frequently asked on a particular news group and the acknowledged best answers.

What was almost entirely missing from the mass of topic-based tech comm that grew up on the Web was any contribution from anyone with the job title Technical Writer. While users, mavens, and engineers were writing thousands of technical Every Page is Page One topics on the Web, official documentation tended to remain in the form of books and local help systems.

Today, more and more tech writers are publishing content to the Web, but often in the form of books formatted as PDF or as help systems, rather than in the form of content conceived and created to work on the Web. In recent years, a growing interest in wikis has changed this picture somewhat. Wikis are a natural Every Page is Page One medium, and some companies are now using wikis for their technical documentation. Companies that were born on the Web have also tended to produce Web-based documentation from the start (and often exclusively), and their growing numbers are helping to tip the scale. Today, therefore, there is considerable movement of professional tech comm into writing for the Web as a hypertext medium. As we move in this direction, we need to think of the topic as a unit of presentation, and of every topic as a potential page one for any reader.

Every page is still page one even if the reader reads several

Saying that every page is page one is not saying that the page is the first page the reader has read today, nor that the reader will only read one page to complete one task. What it means is that every time a reader comes to a new topic, that topic operates as a new page one, just as when a reader puts down one book and picks up another, the first page of that new book is a new page one. Because a reader's course is self directed, there can be no writer-designated page two or three. Readers skip about and accept that there is a reset between each topic. And no matter how many topics a reader encounters, every one of those pages is a new page one.

Characteristics of EPPO topics

The following chapters will detail the principal characteristics of Every Page is Page One topics. Here is a quick overview to set the stage:

- **Self-contained:** An EPPO topic is self-contained. It has no previous topic and no next topic. It does, however, rely on the whole information environment in which it is located for supporting and ancillary information.
- **Specific and limited purpose:** An EPPO topic has a specific and well-defined purpose. This is highly related to the purpose of the person who is reading it, but it is not the same thing. One topic has to serve many readers, and is designed to serve a community, not an individual.
- **Conform to type:** It turns out that, unlike book length content, Every Page is Page One topics seem to naturally conform to fairly well-defined types, often the result of a community process that develops the best way to treat a particular kind of subject. The type of a topic is based on its purpose: the type defines the information necessary to serve its purpose.
- **Establish context:** Readers can come to an Every Page is Page One topic from anywhere. An EPPO topic must establish its context in the real world so readers knows where they are and what to expect.
- **Assume the reader is qualified:** An EPPO topic assumes readers are qualified to complete the specific and limited purpose of a topic. Readers who are not fully qualified can read other topics to get the information they need.
- **Stay on one level:** Books tend to change their level of abstraction and detail over the course of the narrative. But information-foraging readers prefer to choose for themselves whether to go for detail or the big picture. An Every Page is Page One topic stays on one level and allows readers to change levels whenever they wish by changing topics.
- **Link richly:** An EPPO topic is meant to support effective information foraging. Therefore, it links richly along the lines of subject affinity to help the reader follow the scent of information.

CHAPTER 7

EPPO Topics are Self-contained

When asked to describe what a topic is, almost everyone in the business will use the same adjective: "stand-alone." But, as we saw in Chapter 6, what stand-alone means depends on what you mean by topic. For Every Page is Page One topics, stand-alone really means self-contained.

Let's look at an example. One of the most obvious is a recipe, like the Tarragon Mac and Cheese recipe in Example 7.1.

There are several parts to this recipe: a title, introduction, picture, list of ingredients, directions, number served, and notes. Each of these can stand alone in the building block sense. The introduction, for instance, is grammatically complete (you can read it and understand what it says) and structurally complete (it is a whole paragraph). But by itself, it is not sufficient for action. You can't cook Tarragon Mac and Cheese based on the introduction.

To function alone, the recipe needs all its parts. To be certain, some of the parts can be considered optional. The recipe does not need the picture to function, any more than your car needs heated seats to function. But the optional pieces, like the picture and the introduction, make the function more pleasant.

If you are familiar with DITA and its notion of concept, task, and reference topics, you may look at this recipe and see several topics. Not every DITA practitioner I have talked to agrees on how to divide a recipe into DITA topics; some might argue for keeping it as a single topic. However, a typical approach would be to treat the introduction as a concept topic, the ingredients list as a reference topic, and the instructions as a task topic. Such topics would then be the building blocks of a recipe – in other words, they would be building-block topics.

Example 7.1 – Tarragon Mac and Cheese Recipe

Tarragon Mac and Cheese

After eating some tasting menu or another (I do so many, I can hardly keep track), I wanted to replicate a taste combination of garlic and tarragon. This is not an uncommon combination by any stretch of the imagination, but it was after this tasting that I was making the Good Eats baked macaroni and cheese that I decided to make this variant. The tarragon adds a lovely undertone of sweetness that takes what would otherwise be a fantastic macaroni and cheese and turns it into something slightly exotic.

Ingredients:

- ½ lb. elbow macaroni
- 3 tablespoons butter
- 3 tablespoons flour
- 1 tablespoon powdered mustard
- 1 tablespoon garlic powder
- 3 cups milk
- ½ cup yellow onion, diced
- ½ teaspoon Tarragon, Fresh or Dried
- 1 large egg
- 6 ounces extra sharp cheddar, shredded
- 10 ounces Colby, shredded
- 1 teaspoon kosher salt
- Fresh black pepper

Topping:

- 3 tablespoons butter
- 1 cup panko bread crumbs

Directions:

Oven to 350°F.*

Cook your pasta. Remember, it's going to bake some more, so leave a little bite there.

Melt the topping butter in a pan and mix in the panko. Set aside.

Take the 3 tablespoons of butter and melt in a large sauce pan. Add the onions and sweat (or sofrito). Whisk in the flour and stir for a few minutes, until there is a nutty smell or until the flour starts turning a shade or so darker of brown. Stir in the milk, herbs, spices, and salt. Simmer for 10 minutes. Taste and add more salt if the béchamel is lackluster in flavor.

Crack the egg into a small bowl and pour in a bit of the béchamel to temper the egg. Mix that, and pour it into the sauce pan. Stir in 2/3 of the cheese. Stir or fold the macaroni into the cheese sauce and pour into a 9 x 12 baking dish, or a deep 8" round casserole, or whatever seems to hold it best. I tend to use the pyrex baking dishes because they have a convenient cover and carrying case for taking to parties.

Cover with the rest of the cheese and cover that with the buttery panko.

Bake for 30 minutes. If, for whatever reason, the panko is not golden brown and delicious, put it back in until it is.

If you are of strong will, let rest for a few minutes before eating. I usually do that. For the second serving.

(Serves 6)

—"The Food Geek"[11]

Whether a building-block approach is the best way to manage your content is outside the scope of this book. However, there's no question that presenting the introduction, ingredients, and instructions separately is not useful to the reader. If you are going to manage those elements separately, for whatever purpose, you need to bring them back together before you present them to the reader.

Similarly, we are not interested here in whether the topic, or any of its constituent parts, is suitable for reuse. Here we are only concerned with whether the topic (the recipe as a whole) is *usable*, not whether it is *reusable*. (I will have more to say on the subject of reuse and EPPO topics in Chapter 21.)

Self-contained, not all alone

To say that a topic is self-contained is to say that it is not designed to work only as part of some larger information product. But neither is a topic expected to work in a complete information vacuum. Indeed, many of the topics you find online are useful precisely because you can highlight a term or concept you don't understand and select `Search Google for...` to find more information.

Suppose you are making the Tarragon Mac and Cheese recipe (Example 7.1), and you don't know how to cook pasta. You search for "cook pasta," and you get lots of help. Cooking pasta is a generic process, and you can learn it from many different places. The recipe does not depend on any one specific topic to ensure that the reader can learn to cook pasta. It relies on the whole information environment in which cooking occurs. Cooking is now an activity that takes place in the context of the Web.

Therefore a topic is self-contained not because it is entirely self-sufficient, but because it exists in a rich information environment that readers can call on to further their understanding.

The information scent of self-contained topics

Good information scent improves findability. Making sure your topics are self-contained will help give them the right scent.

There is nothing worse then following the scent of pizza into the lunch room only to find nothing left but crust. The scent of pizza hangs about the place. It even says pizza on the box, but there is no pizza here. Just crust. You leave frustrated and hungry.

Making sure your topics are self-contained will help give them the right scent.

This can happen when you search for technical content on the Web or in a captive help system. At first a hit may look good, but when you click through you find just a couple of paragraphs of introductory matter or an out-of-context procedure that you can't be sure applies to your situation.

Often what you have found is a fragment of a book that has been burst to create a help system or a building-block topic standing by itself. The material you are looking for may be close by, so you could try the back and next links, but if you have hit a chapter introduction, the meat of the information could be many clicks away, and if you have an out-of-context procedure, you might have to click back several times to find the context.

A good EPPO topic that is self-contained represents a complete meal for a hungry information seeker.

CHAPTER 8

EPPO Topics have a Specific and Limited Purpose

If a topic is going to be self-contained, we must be able to set definite boundaries around it. We can only determine if a topic is self-contained if we know what purpose it is meant to fulfill. A topic needs a specific purpose.

The scope of a topic

A topic is sometimes described as something that answers one question. But an answer can be too simple and granular. "42" and "Paris" are the answers to single questions, but they are not useful topics. And if a topic answers more than one question, must it be subdivided until it contains only one answer, no matter how small? Would the result be a useful topic? What if completing a task requires answering more than one question? How many questions does the Tarragon Mac and Cheese recipe answer?

Tom Johnson examined the difficulty with using questions as a guide to topic size in his blog post "Why Long Topics Are Better for the User":[1]

> What's a *good* question? Questions can scale to a low or high order, being very specific and mundane to being abstract and conceptual. A one-sentence topic might provide the answer to a question (e.g., How many feet are in a yard?), while a 300-page dissertation might provide the answer to another question (e.g., What was Chaucer's influence on the Renaissance?).
>
> In other words, you could construct the question so that it scales for any length of topic.
>
> However, if you can construct an intriguing question (or at least a relevant question within the user's business scenario), that question merits enough information for a good-sized topic.

[1] http://idratherbewriting.com/2013/05/06/why-long-topics-are-better-for-the-user/

Tom seeks to solve the problem by distinguishing good questions from bad. But what is a good question? Tom suggests that it is "a relevant question within the user's business scenario." The user's business scenario is the purpose that the user is attempting to achieve.

Purpose describes a unit of work. If you ask people what their purpose is, you will get an answer on a human scale. Anything below that level is probably too granular.

You can see the inadequacy of questions as the defining scope of a topic by visiting Q&A sites. You will find many cases where the questioner has asked a vague and general question, or a highly specific but uncontextualized question, and one or more respondents have written back saying something such as "What are you trying to do?" Before providing an answer, they need to understand the questioner's purpose. A question can only be answered properly in the context of a specific purpose.[2]

Task-based writing

Writing a topic to serve a purpose for the reader is task-based writing. Task-based writing is commonly defined by contrasting it to feature-based writing: "Describe the user's task, not the product's features." This can lead writers to think they should not mention a product's features at all or that they should never express user tasks in terms of product features. While that may sound noble and grand when expressed as a generality, writers quickly find that it is impossible to do in practice.

I believe this confusion arises from a failure to distinguish between *motive* and *purpose*. The motive is the reason someone wants to complete a task. The purpose is that person's plan to complete the task and satisfy the motive. We are a tool-using species. When we create a plan of action, it generally involves the use of tools. Readers very seldom come to technical documentation, or Google, asking abstract questions. They come looking for specific information on how to implement their plans, and they frequently express those queries in terms of their tools.

---

[2] Of course, some questions are purely requests for data, such as asking for the current time. These can be answered without regard to your purpose for asking. But, because such answers are purely data points, they are not the subject of topics.

Tools don't stand apart from our tasks. Rather, tools express a vision or a method for how a job should be done, and as we use a tool regularly, we incorporate that tool and its methods into how we think about the task itself.

When John Carroll looked for ways to make it easier for readers to grasp word processing concepts, he re-phrased the instructions in terms his readers already knew, including topic titles such as "Typing something"[8, p. 112]. Note that he did not use the abstract, tool-free term "Writing something." Instead, he used a term specific to the tool his readers were used to: the typewriter. His readers did not think of their tasks in terms of abstract ends, but in terms of the tools and processes used to accomplish those ends.

Graphical computer interfaces have exploited this from the beginning, with broad desktop metaphors and more specific metaphors based on physical tools such as files, clipboards, and scissors. This language has now become the language of computing, more or less divorced from the physical tools on which they were based.

In short, you can't talk about the user's task without talking about the user's tools.

This is one reason documenting a new tool is such a challenge. Our tools shape our understanding of our tasks to such a degree that it is difficult to separate a task from the tool we currently use to accomplish that task. Anyone who has responded to a technology RFP (Request for Proposal) has probably been frustrated by business requirements written in terms of a client's existing tools and the processes that have grown up around those tools. It can be difficult to make the case for a tool that removes responsibilities or simplifies processes when the terms of the RFP require support for those responsibilities and processes.

You can't talk about the user's task without talking about the user's tools.

One of the hardest things about moving technical writers from desktop publishing to structured writing is persuading them to give up responsibility for how the final output looks. Writers will keep looking for ways to specify layout, even in markup languages specifically designed to remove layout concerns. They understand their jobs in terms of the responsibilities their old tools imposed on them.

Users express their purposes in the context of their tools and processes. As they learn more about their tools, their queries become more specific and tool-focused, and when they change tools, their language becomes more that of the new tools and less that of their old tools. But even as tools change, users do not separate their purpose from their tools.

Users do not separate their purpose from their tools.

Derived purpose

Just as readers do not always express their queries in terms of their original motive, neither do they always express them in terms of their overall purpose. In many cases they express their queries in terms of what we might call a *derived purpose.*

There is a whole sub-genre of adventure stories that consists of a quest to assemble the parts of a key. The princess has been locked in a castle. The key that opens the castle gate has been broken and the pieces scattered to the ends of the earth. The hero's motive is to marry the princess. His purpose, in pursuit of this end, is to unlock the castle gate. To achieve this purpose he must first find each piece of the key. The quest for each piece becomes a separate story in its own right – a derived purpose.

How the derived purpose is expressed is often determined by the larger story. In the prolog of the series, the prince learns from a wise old man about the locking of the gate and the scattering of the keys. The prince then describes his first quest as "finding the key in the west." The prince has learned the names of things in his main quest, and he uses those names in his subsequent derived quests.

Similarly, a reader will often come to the documentation with a derived purpose stated in terms of the features of your product, because he or she encountered those features while pursuing the broader purpose.

Therefore, purpose cannot be divorced from features. What distinguishes task-orientation from feature-orientation in documentation is not what you call things, but what you choose to say about them. Feature-oriented documentation contains known facts about a feature, regardless of their utility. Task-oriented documentation contains information on features that helps users accomplish tasks.

Defining the purpose of a topic

In Example 7.1, "Tarragon Mac and Cheese Recipe," the specific and limited purpose is to show an experienced cook how to prepare Tarragon Mac and Cheese. The purpose is specific and strictly limited: cook this one dish. It does not teach basic cooking techniques. If you want to learn how to cook elbow macaroni, you can learn how elsewhere. The recipe does not give you the history of cheese or the evolutionary biology of tarragon. It is about cooking Tarragon Mac and Cheese and that's all.

Example 8.1 is the outline of the topic "Using Themes from the WordPress Codex."[3] This is an Every Page is Page One topic, and its purpose is to enable the reader to use WordPress themes. This is, of course, a derived purpose. The reader's motive is probably something like "sell more products," and the purpose is to make this website more attractive and functional. At some point this person has learned that you can use themes to make a WordPress site look the way you want it to (perhaps from a topic on making a website more attractive and functional), and that knowledge leads to the derived purpose of using a new theme on the site.

Example 8.1 – Outline of Using Themes for WordPress Codex

Using Themes
- What is a theme?
- Get new themes
- Using themes
- Adding New Themes
  - Adding New Themes using the Administration Panel
  - Adding New Themes by using cPanel
  - Adding New Themes Manually (FTP)
- Selecting the Active Theme
- Creating Themes

Like the Tarragon Mac and Cheese recipe, this topic is self-contained in the sense that it functions alone. You don't have to read a particular topic before or after you

[3] http://codex.wordpress.org/Using_Themes

read it. If you search for "WordPress themes," you might land directly on this topic without going through the front door of the WordPress Codex, and you may not even be aware that you are in the Codex at all.

Note that this topic contains no less than four procedures, including three alternate procedures for adding a theme. If you want to use a theme on your site, choosing among those three alternatives is part of one task. In addition, to use a theme you must first install it and then activate it. Presenting these procedures in separate topics would not help you complete the task. Placing the four procedures into separate source files might help the writer maintain the content set, but a topic consisting solely of a procedure for installing a theme manually, without procedures for finding and activating it, probably would not be functional for most users.

But if the topic does all these things, does it meet the condition of having a specific and limited purpose? Yes it does. Its purpose is to enable the reader to use WordPress themes. It contains the key information that a user would need to find, install, and activate a theme. That is a reasonable purpose. It corresponds to a real job that a typical user would want to do.

Both the Tarragon Mac and Cheese recipe and the Using Themes topic work within clear limits. Both mention ancillary subjects that might be of interest to some readers, and the Using Themes topic is particularly generous in providing links to those subjects (cPanel, FTP clients, the Administration panel, etc.). But neither topic strays from its purpose. Each does its job and leaves the rest for others.

Topic purpose vs. user purpose

I talked earlier about the common misconception that a user's purpose should not be stated in terms of product features. Now I must deal with a related misconception, which is that a topic's purpose is synonymous with the reader's purpose. The purpose of a topic is to serve the purpose of a reader. However, that does not mean that a single topic is a personalized expression of a particular individual reader's entire purpose.

A taxi will pick you up at your door and drop you off at your destination, serving you alone and ready to take any detour that interests you. That is what every writer would love to do for every individual reader. In practice, though, we can't.

For products that have a few simple and discrete functions that are used in isolation, you might be able to provide a set of comprehensive topics that come close to this ideal. However, most products have a broad enough set of features and users that it's impossible to document the full set of permutations.

In these cases you need general topics that serve the overall purposes of many users. That is, your topics will need to work more like a bus service than a taxi service. They need to pick your readers up at a logical starting place and drop them off at a logical ending place, but they don't need to go point-to-point from each reader's home to each specific destination.

Every Page is Page One topics serve many readers, and they pick the reader up at a sensible place. They leave it to the reader to get to the departure point. They assume that the reader is *qualified* – that is, either ready to perform the task immediately or able to acquire any pre-requisite knowledge. However, they offer transfers and other connections via links (see Chapter 13).

A collection of related topics can provide an efficient transportation network that enables many different readers to accomplish tasks while sharing each leg of the journey with other readers.

A topic, then, can serve a purpose for many readers, but that purpose is not necessarily customized to each reader's specific purpose on every occasion. It may be tempting to imagine creating topics for every subtle variation of every user's purpose, but beyond the obvious expense, you will find that search engines won't distinguish your variations very well and users may not look past the first close-enough topic anyway.

A well-designed information set is like a well-designed transportation system, it allows passengers to travel individual itineraries along shared routes.

Purpose and topic size

One of the most frequent questions writers ask is: how long should a topic be? Over the years I've seen that when writers are first introduced to the idea of topic-based writing, they almost always tend to produce topics that are too small to be useful. Once they discover how easy it is to cleave one piece of text from another, writers gleefully chop their content more and more finely.

The result is a sea of tiny fragments that may qualify as topics – they are grammatically complete and convey one idea – but which are impossible for a reader to navigate or glean any meaning from. When you reach that point, there are three ways out:

1. Give up and go back to writing books.
2. Create a way to string fragments together into something larger.
3. Reset your idea of what a topic is and come up with guidelines for creating topics that are correctly sized and structured to be useful to the reader.

The first two options present less of a challenge to the ingrained habits of book-based information design, and one or the other is often chosen. This book, of course, is about the third option.

Give readers not only the action, but the reason and context for acting.

The key to finding the right size for an EPPO topic is to define the purpose correctly and then write a topic that fulfills that purpose. A focus on purpose forces you to scale each topic to a real need, giving the reader not only the action, but the reason and context for acting.

Decision support and the reader's purpose

Providing the reason and context for acting is really another way of saying, "provide support for decision making." One of the most important tools of modern business is the decision support system. Such systems can be complex, but at their heart, they simply provide people with the information they need to make decisions.

A task is not a procedure

In technical communication, we don't talk much about decision support; we talk about task support. We frame our jobs as providing the information people need to complete their tasks. Unfortunately, the information we provide is often simply a procedure for operating a machine. A task is not a procedure (a theme I'll return to in Chapter 9). In many cases, the information people need to complete their tasks is not information on how to operate machines, but information to support their decision making. It's not "how do I push the button," but "when and why should I push the button and what happens if I do."

In her recent article on TechWhirl, "Tips and Tricks: Getting from Obvious to Valuable Technical Content"[3], Ena Arel talks about the kinds of questions she finds herself asking when reading documentation:

> As I was reading, I found myself asking "Why do I need to know this?" "What does this term really mean, and how does the corresponding concept impact my product use?" "Why did you use '4' in this example? Is this the value I should use too? How do I decide what value I need to use?" "These results you are showing me, are they good? Why or why not?"
> — "Tips and Tricks: Getting from Obvious to Valuable Technical Content"[3]

None of these questions is about how to push the buttons. They are all about decision making. Tom Johnson says much the same thing in his blog post "Misconceptions about Topic-Based Authoring":

> The real heart of technical instruction doesn't lie in the step-by-step how-to information. It lies in understanding concepts and how they work together to produce an end. This focus on the conceptual interplay of the parts should drive the technical writing experience, both from a reader and writer's point of view. Procedures are more like footnotes. As soon as the user understands the *why* and the *what* and the *who* and the *where*, the *how* is merely a mundane detail.
> —"Misconceptions about Topic-Based Authoring"[4]

Too much documentation covers only the physical procedure, leaving out any help in making the decisions necessary to complete the task.

[4] http://idratherbewriting.com/2012/07/31/misconceptions-about-topic-based-authoring/

I'm not saying that you never need to document the physical procedure. I have spent years writing for software developers, and I know that for many applications, the details of command syntax have to be clearly documented. The same thing holds for other forms of documentation, too. But simply documenting procedures is never enough. Supporting the decisions users need to make, large and small, is the tough part.

To be clear, I am not talking about telling users *what* their decision should be. That is specific to each situation. What I am talking about is documenting the context, letting users know what decisions they must make, making them aware of the consequences, and, as far as possible, leading them to resources and references that will assist them in deciding what to do. I'm talking about answering questions like:

- Where are the valid values for this field listed?
- What do each of the field values mean?
- How will the system change as a result of this setting?
- Does this setting form part of a collection of settings that are used to achieve an overall objective for the system.
- What are the side effects of setting a particular value? Are there trade-offs on performance, access, or security as a result of changing this setting?
- Should this setting match a value set elsewhere in the system? If so, which value, and which is the master and which is the slave?
- Are there larger questions to consider before choosing the value for this setting?
- Will the system validate this setting? How will I know if I have the right setting?
- Does my choice for this setting depend on what other users have done, and, if so, what questions do I need to ask them before I change this setting?
- Can I change this setting later, or will there be irrevocable consequences?
- Could this setting result in loss of data or change how data is processed?
- Who else might be affected by this setting, and what do I need to tell them so they can make good decisions about their own parts of the system?
- How is this setting affected by optional components?

A good Every Page is Page One task topic should address these kinds of questions and should link richly to ancillary material the reader may need to help answer these questions. Only when the planning and decision-making aspects of the user's task have be thoroughly covered should the topic proceed to the physical procedure for executing the decisions the user has made.

Technical documentation is a decision support system. Insofar as it fails to support the user's decision making, it fails its purpose, even if all the physical procedural steps are documented correctly.

Purpose and findability

Keeping a topic to a single purpose is a huge aid to findability. People usually have a specific and limited purpose when they search. As Gerry McGovern notes:

> [W]hen was the last time you went to Google and searched for "anything interesting out there" or "I'm bored. Show me something new."?
>
> The vast majority of people who go online know what they want.
> —"Communications and marketing professionals at a crossroads"[5]

McGovern may be overstating this. People often go to the Web because they are bored and want distraction. He's right about the search terms, though. People may go to sites like Tumblr, Buzzfeed, Facebook, Twitter, and Cheezeburger to pass the time, but by the time they follow a link from those sites to your content, a purpose has formed. Something specific has piqued their interest, a scent of intriguing information, and you had better deliver what that scent has promised.

When people search the Web, they are looking for content that meets their specific and limited purpose. A good EPPO topic that meets that specific and limited purpose will give them what they want. And because it is specifically written to that specific and limited purpose, it will *smell* like what they are looking for. And because it does what it claims to do, it will be filtered in by search engines and social curators.

[5] http://gerrymcgovern.com/new-thinking/communications-and-marketing-professionals-crossroads

CHAPTER 9

EPPO Topics Conform to a Type

A topic type is a plan or a prescription for a topic. It tells the writer how the topic should be written and the reader how it should be read. A topic type defines the content, order, and form of a topic.

As we saw in Chapter 7, a topic (a recipe) whose purpose is to enable an experienced cook to prepare a particular dish is almost always written with the same core content, in the same order, and with the same form for each part.

- The core content is the name of the dish (title), a list of ingredients, the preparation steps, and some standard information items like prep time and serving numbers. It also includes optional items like introduction, wine pairing, or a photograph of the dish.
- The order is an optional introduction followed by a list of ingredients, the directions, and other optional information. If there is a photograph, it is almost always in a lead position under the title.
- The form for the list of ingredients is a list with one ingredient per line with the ingredient name to the left and the quantity and unit of measure to the right. The form for the procedure is a set of numbered steps. Other elements, like servings or wine pairing, are generally presented as key value pairs separated by colons.

The content, order, and form described here are specific to the recipe topic type and are based on the purpose recipes are designed to serve. You would not use the same content, order, and form for other purposes.

Good Every Page is Page One topics frequently share a clear topic type with other topics that have a similar purpose. We often think of a topic type as something imposed artificially on the content. Yet millions of EPPO topics, written by people who have no knowledge about topic type classifications, fall into recognizable topic types without any imposed structure.

This is certainly the case with recipes. When you write down a recipe, you don't have to spend any time wondering what to say or how to organize the information. You know the topic type of a recipe. You know what the required fields are and what optional fields you have to choose from.

Conforming to a type is the principal way in which we ensure that an EPPO topic meets its specific and limited purpose. But more than that, conformance to a type helps the content smell right. A recipe or an API reference that follows the established pattern for recipes or API references looks right and reassures readers that they have found what they are looking for.

A recipe or an API reference could be written to contain the same information without following the conventions for its type, but then it would not look or smell like a recipe or an API reference. Readers might arrive at the topic and not recognize that it contains the information they want, simply because it doesn't look like what they expected.[1]

A good technical communication example is the API reference topic. Example 9.1 shows a typical API reference topic from the *Chrome API Reference.*[2]

This isn't the most extensive API reference entry you will see (I chose it for brevity rather than completeness), but it conforms to a type you will find in almost any function API reference. The standard API reference type looks like this:

- Function name
- Function prototype (which itself conforms to a well defined structure)
- Return value (if there is one)
- Description
- List of parameters and their descriptions

[1] This phenomena is explained in Detection Theory, which deals with how people or systems distinguish signal from noise. http://en.wikipedia.org/wiki/Detection_theory

[2] http://developer.chrome.com/apps/runtime.html#method-sendMessage

Example 9.1 – Chrome API Send Message API example

sendMessage

```
chrome.runtime.sendMessage (string extensionId,
                            any message,
                            function responseCallback)
```

Sends a single message to onMessage event listeners within the extension (or another extension/app). Similar to chrome.runtime.connect, but only sends a single message with an optional response. The onMessage event is fired in each extension page of the extension. Note that extensions cannot send messages to content scripts using this method. To send messages to content scripts, use tabs.sendMessage.

Parameters

extensionId (optional string)	The extension ID of the extension you want to connect to. If omitted, default is your own extension.
message (any)	
responseCallback (optional function)	

Callback

If you specify the *responseCallback* parameter, it should specify a function that looks like this:

```
function ( any response ){...} ;
```

response (any)	The JSON response object sent by the handler of the message. If an error occurs while connecting to the extension, the callback will be called with no arguments and lastError will be set to the error message.

You will find variations on this type. For example, the order of the fields may differ from one reference to another and there may be additional information, but the basic form of an API reference is clear in each case.

This is not an accident. The reason API reference entries follow a consistent topic type is because they all serve the same specific and limited purpose they enable programmers to use the API correctly. Form follows function. Type follows purpose.

The evolution of topic types

Patterns help writers create topics and help readers recognize and consume topics more quickly.

Why do so many topic types follow common patterns? Because common patterns help writers create complete topics and help readers recognize and consume topics more quickly. Topics that conform to a type are more likely to be filtered in to a user's search – they have a better information scent and are more likely to perform effectively for the reader. Thus the successful types rise to the top and become the exemplars that other authors follow.

It is not surprising that recipes conform to a type. But it is worth noting that the recipe type was not created by a standards group. It arose naturally out of the experience of millions of cooks writing millions of recipes over centuries of civilization.

A used car review is another example of a common topic type. A used car review may look like just a sequence of paragraphs. But if you look more carefully you will see that most used car reviews cover the same basic things:

- overview
- equipment
- notable features
- interior and comfort
- safety
- economy
- reliability
- price history

The order may differ, but pretty much every used car review will include the same basic information.

Once again, this type is not the product of a standards committee. Rather, it comes from the needs of used car shoppers. Reviews that provide what shoppers need to know rise to the top and influence other reviewers.

Used car reviews would probably be easier to read if each section was called out with a consistent header. But don't fall into the trap of assuming that every element of a topic type is necessarily visually set apart from the other elements. Topic type is determined by the information needed to fulfill a purpose, not by its visual appearance. A visual layout that reflects the type usually helps the reader, but it is the type that is the master and the layout the slave, not the other way round.

When a topic has a specific and limited purpose, it is likely that all topics that serve a similar purpose will contain the same kinds of information. Because a used car review serves a specific limited purpose for a used car shopper, it naturally covers the standard set of information a used car shopper needs and, thus, conforms naturally to a consistent type.

Wikipedia is a great place to look for topic types. For instance, the Wikipedia topic for nearly any city conforms to a well-defined type. A table of contents sidebar such as the one in Figure 9.1, "Wikipedia city topic," which is for the entry for Ottawa, lays out the major divisions of the topics: History, Geography, Education, Economy, Culture, and so forth. The entries for other cities follow a similar structure.

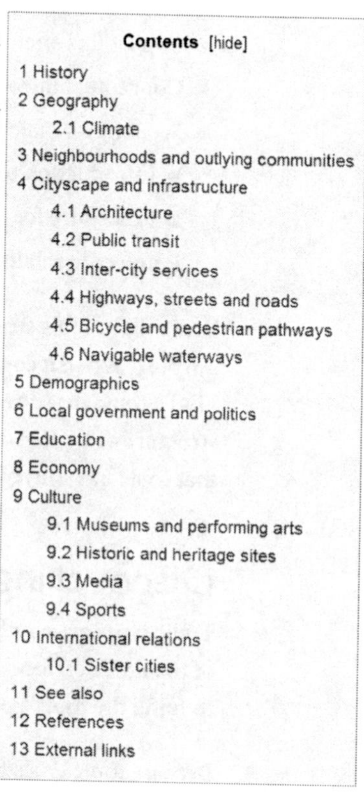

Figure 9.1 – Wikipedia city topic

Many other topics in Wikipedia have similarly well-defined topic types: vehicles, languages, flora, fauna, novels, and on and on. Simply browsing Wikipedia is an effective short course on topic typing. Once again, no standards committee established these topic types. Rather, they are the result of thousands of contributors gradually building up topics, filling in gaps, and refactoring and refining the structure.

Even topics such as the WordPress Codex topic on Using Themes (Example 8.1), which at first glance may not seem to conform to a type, often do have a fairly consistent structure. A typical structure for this topic type looks like the following:

- Topic title (often with a consistent gerund-noun structure)
- Conceptual introduction
- Sections on major tasks (getting, using, creating, in this case)
  - Introduction to the action
  - Procedure for the action
- Pointers to related information

You will find this structure, with minor variations, in many topics with a similar purpose. When it comes to topics, adherence to a type is the norm, not the exception. Find a topic that does not seem to have a type or does not conform to the common structure of topics with a similar purpose, and you will almost always discover that that topic has strayed from its purpose or never had a well-defined purpose.

Discovering and defining topic types

Topic types are a reflection and a formalization of the specific and limited purpose of topics. Therefore, defining explicit topic types begins by exploring what is needed to fulfill the topic's purpose.

Because topic types fall naturally out of the specific and limited purpose of a topic, you might think your topics will naturally fall into types without any attempt to explicitly think about the topic type. Unfortunately, it's not that easy. While good Every Page is Page One topics do naturally conform to specific types, there are many bad topics out there as well. Using well-defined, explicit topic types will help you create good EPPO topics, and also make it easier for a team to collaborate on a consistent set of EPPO topics.

Creating topic types is a two-part effort of discovering existing topic types and then using your discoveries to define the topic types you need. Once you do that, you will want to document your topic types and set up your authoring environment to support them. The best way to do that is through structured writing, which I will cover in Chapter 18.

Discovering topic types

One of the most effective ways to discover topic types is by looking at existing topics designed to serve the same purpose. You can look on the Web, in your competitor's documentation, and in your own existing documentation. Make a collection of diverse sources and make lists of the repeating fields and sections that you find in each sample.

Remember when you do this that you are not looking to define a generic template that would fit all of these examples as they currently exist. You are looking for something much more specific: the information fields that are required to fulfill the specific and limited purpose of your topic type. The words *specific* and *limited* are key here. You are looking for the limited set of specific pieces of information required to meet the user need this topic is designed to fulfill.

Defining topic types

Once you have done your research, it is time to create your topic definition. Now is the time to get specific. A recipe doesn't contain just any list, it specifically contains a list of ingredients. And each entry in a list of ingredients has a specific form that looks something like the following:

```
[ingredient name]............................... [quantity] [unit]
```

An API reference doesn't just contain a line of code. It contains a function signature, and that function signature has a particular format that all programmers understand:

```
[return type]? [function name] [[parameter name] [parameter type]]...
```

Even when topics naturally conform to a type, individual authors may implement that type with a different organization and different inclusions and exclusions. For a systematic authoring project, you need create a stricter definition of each topic type to ensure consistency and completeness.

If you use a structured methodology (see Chapter 18), you can encode your topic type in a database or using an XML schema. However, the way you encode your topic type, while important, is an implementation detail. What is most important is to make sure you capture the information a topic needs to have to serve its purpose.

To do this, you need to start with the specific and limited purpose you have defined for each topic type. What information must each topic include to achieve its specific and limited purpose? What information does a user need?

Stay focused on the specific and limited purpose. It is easy to start imagining all kinds of things a hypothetical user might want to know. By all means keep a list of these things, because they may be clues to other topic types you need, but don't let them creep into your topic type definition. The readers of a recipe do not need to understand the evolution of tarragon, the history of cheese, or the manufacturing of macaroni in order to make Tarragon Mac and Cheese.

Handling optional material

You can have optional parts in your topic type, but you should only include them if they are sufficiently related to the topic's purpose. For instance, a wine pairing suggestion for a recipe is not something every reader needs, but it relates directly to the purpose of the recipe, which is to produce an enjoyable dining experience. The evolutionary history of tarragon, on the other hand, though it may be of interest to a few readers, does nothing to enhance the pleasure a diner takes in the meal and, therefore, is unrelated to the topic's purpose.

This is not to say that your content should never contain this kind of decorative information. Many recipes include anecdotes that can entertain even if they don't contribute to the preparation of the dish. However, such entertainments can become tedious if indulged and may be inappropriate in a business setting.

Serving the commercial purpose

In addition to the needs of the reader, you need to serve your commercial needs. A company produces content to capture the reader's attention and money. If you are considering adding decorative elements to your topic type definition, you need to consider your company's commercial purpose. Do the decorative elements help attract and hold customers? Can you prove that they do, or are you just guessing? Ultimately, every part of a topic type definition needs to serve either the reader, the publisher, or (preferably) both.

Concept, task, and reference reconsidered

It has become an axiom of technical communications in the last few years that all content falls into one of three types: concept, task, or reference. However, EPPO topic types are more specific and varied than this simple trio. To say that every technical communication topic must be a concept, task, or reference is like saying that every object must be an animal, vegetable, or mineral, as in the game 20 Questions.

The question is not so much whether this broad characterization is true, as what it is useful for. If you are duck hunting, you want a Labrador Retriever, not a hyena. If you are competing in the Monaco Grand Prix, you want an F1 race car, not a pickup truck. If you are making a soufflé, you need eggs, not turnips. Animal, vegetable, and mineral are not sufficiently precise for these purposes. Similarly, a recipe, the procedure for fueling an ICBM, or the instructions for knitting a sweater are all tasks, but they differ significantly in content and structure.

We could simply choose to regard concept, task, and reference as generic building-block topic types and, therefore, as orthogonal to EPPO types.[5] However, there is a definite popular view in technical communications at the moment that holds that all topics presented to the user should be assigned to one of these three generic types. Therefore it is important to address why more precise topic types are needed.

The type of an Every Page is Page One topic is defined and limited by its purpose. If the topic is a recipe, you need the parts of a recipe, not the parts of a sonnet or an API definition. Therefore, the type of the Tarragon Mac and Cheese topic is recipe, not

[5] You may wonder why I don't address DITA specialization here. Specialization is a DITA mechanism that lets you create a more specific topic type from one of the three base types. While you can use specialization to define EPPO topic types, some DITA practitioners would object, saying that to maximize reusability you should build EPPO topics from smaller building blocks rather than create stand-alone topic types.

In any case, the main reason for not talking about specialization here is that even with specialization, the commonly held notion is that all topics are either concept, task, or reference.

task. Recipe topics may belong to the general category of task, in the same way that a hyena and a Labrador Retriever belong to the general category of animal, but if you are writing an Every Page is Page One topic on how to prepare Tarragon Mac and Cheese, you need the more specific recipe type.

The origins of concept, task, and reference

The concept, task, reference trio originates with DITA's adoption of these three types, which are a reduction of Information Mapping's six information block types:

- Principle
- Process
- Procedure
- Concept
- Fact
- Structure

Over the years, other lists of information types have been proposed, but, except within Information Mapping's walled garden, these have all faded away in favor of the trio of concept, task, and reference.

The problem is that, in the popular conception, the words concept, task, and reference have been reduced to shapes: a reference is a table, a task is a procedure, a concept is ordinary text. Figure 9.2, from Tom Johnson's blog post "Unconscious Meaning Suggested from the Structure and Shape of Help,"[6] illustrates this nicely.

We have somehow gone from the laudable idea that users want information that helps them perform a specific task (as opposed to information that simply described the machine) to presenting single procedures by themselves. This is sometimes done in the name of minimalism, despite John Carroll's finding that learners don't follow procedures[8, p. 74].

---

[6] http://idratherbewriting.com/2012/07/18/unconscious-meaning-suggested-from-the-structure-and-shape-of-help/

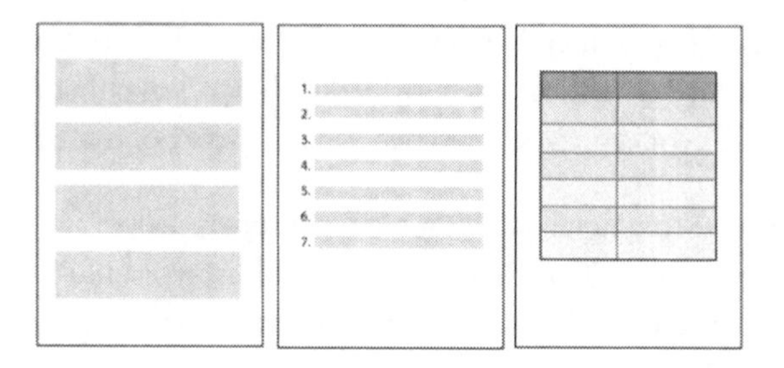

Figure 9.2 – Tom Johnson's "Shapes of Help" graphic

This is certainly not the approach advocated in Information Mapping. IM uses the word *block* rather than *topic* for its six basic information types. Information blocks are not an end in themselves. They are supposed to be combined into maps and presented to the reader as a well-structured document, not as individual information blocks. Information Mapping is as much about how to assemble maps as it is about how to type information blocks.

In DITA, on the other hand, a map is simply a technical device for bundling topics. Beyond the idea that it is useful to put your tables and procedures in separate files, DITA has no information design theory. As the white paper *Information Mapping® and DITA*[7] points out:

> No [writing] principles [are] defined except for the concept of a Topic standing on its own.

And:

> Information Mapping®'s principles provide guidelines to writers to ensure that their content is organized and presented in a useful and effective way for the reader. There is no equivalent in DITA.

---

[7] http://www.informationmapping.com/en/resources-en/whitepapers

This is not a bad thing. It is not bad for a technology to be separated from a design philosophy, even if one is intended to support the other. If there is a problem with DITA, it is not that it lacks a theory of information design, but rather that many people believe that DITA's concept/task/reference trio is a theory of information design.

The result is that when you talk about topic types today, people's minds go at once to the DITA trio. For EPPO, that's a problem because a typical EPPO topic has a much more specific type definition, which may contain several different types of information blocks. It is important, therefore, to spend some time looking at why the trio of concept/task/reference, though useful for some purposes, is not sufficient either as a set of topic types or as a principle of information design.

A task is not a procedure

A DITA task topic type is essentially a procedure. According to the DITA 1.2 spec:

> Tasks are the essential building blocks to provide procedural information. A task information type answers the "How do I?" question by providing precise step-by-step instructions detailing the requirements that must be fulfilled, the actions that must be performed, and the order in which the actions must be performed.
>
> —DITA 1.2 Specification[8]

A task is something a user has to do, a goal to attain. A procedure is a set of instructions for manipulating a machine.

A procedure is, in the Information Mapping sense, an information block. It is a distinct literary form with a defined set of rules and a clear beginning and end. You can certainly create and structure procedures as discrete blocks. But they are not task topics.

A task is something a user has to do, a goal to attain. A procedure is a set of instructions for manipulating a machine. Manipulating a machine is never the user's goal in itself. One of the user's derived goals may be to change the system state, but the manipulations required to change the state are not a goal. A procedure, therefore, may be part of an Every Page is Page One task topic, but it is not a task topic in itself. In fact, some task topics don't need to include any procedures.

[8] http://docs.oasis-open.org/dita/v1.2/os/spec/archSpec/dita_task_topic.html#dita_task_topic

A great deal of the technical documentation I have written over the years has been for programmers, where the product was a programming language, an API, or an operating system. People doing programming tasks use the same tools for every task: editors, debuggers, compilers, etc. No one programming task makes any special use of these tools.

For example, creating rules to process an XML content stream and writing code to control shared access to a serial port are two completely different programming tasks. However, to accomplish either task, programmers need to use editors, debuggers, and compilers in the same way they use those tools for every other programming task. Programmers do not need or want to be told how to type a function or set a breakpoint in every single programming topic. Therefore, programming task topics generally contain no procedures at all.

Most programming task topics are built around code samples, not procedures. Indeed, if you were defining a structured task topic type for a programming task, you would almost certainly make at least one code sample mandatory. But you would not make a procedure mandatory. Procedures are simply not relevant in a typical programming task topic.

A configuration task is another common task type, especially when you are talking about operating systems. Unless you have a GUI configuration tool, there are usually no machine manipulations involved. Configuration is done by writing a text file. It might be an XML file or a file containing a series of configuration commands, but whatever it is, configuring a system has nothing procedural about it. In fact, it is almost all about planning.

Configuring an operating system correctly for a particular device is all about making the right set of trade-offs between functionality, size, speed, and security. It is, as noted in Chapter 8, about decision support. A configuration task topic for any sub-component of a system is mostly a checklist of things to think about: how many communication channels will be required? How much memory will each application need? How much stack space? Just as a code sample is central to a programming task topic, a checklist is central to a configuration task topic. Procedures play little or no role in either one.

For one project, we designed a topic type for a configuration topic for components of a real-time operating system. Because of regulatory requirements, the customer needed to perform expensive line-by-line testing of any configuration and coding changes. Therefore, it was essential to be able to show exactly what the downstream effects of a code or configuration change were and to minimize the need for such changes. Therefore, the topic type was designed with a heavy emphasis on planning and on understanding the exact consequences of planning decisions.

The general topic outline included sections on Understanding (context setting), Planning (which had an internal structure consisting of questions to consider and ways to go about answering them), Configuring (which required a list of inputs, a list of outputs, and a formal flow diagram), Building, and Packaging. This was not a generic configuration topic type. It was specific to the business requirements for a particular product, and we had to formally define all of these pieces to achieve consistent and accurate content for each configuration topic.

Topic typing should always be about making sure that every topic of a particular type does the whole job it is supposed to do.

A reference is more than a topic

A reference is more than a topic, it's a database. Not every reference is a topic or a container of topics. Some contain only simple data fields, and a collection of simple data fields is not a topic – whether it contains a dozen entries or a million.

A database is a place to look something up. What you are looking up may be a simple value, such as a torque value for a particular bolt, or the voltage for a particular circuit. In these cases, the information the reader wants and receives is simply a number, albeit one that has been placed in context.

In other cases, a reference may contain discursive content, and the information the reader receives may be an Every Page is Page One topic. For example, an API reference consists largely of a set of pages, each of which describes one routine. Each page follows the same pattern: a number of standard fields, like return type or arguments, and then a general description of the routine, usually with examples of usage.

Some readers will read through the complete topic, at least the first time, so the individual pages do indeed read like topics. On the other hand, a reader may look at an API reference entry merely to find the return value of a function call, or the type of an argument, and not read any of the rest of the topic. The reference is designed to support more than one kind or query for more than one kind of information.

In the paper world, it has been standard practice to present references as tables. If you have a small reference with a few records consisting of a few simple fields, you can fit it into a table that will fit on one page. The idea of a reference being a topic type is often expressed in terms of such tables. But these are topics only in the sense that they are small collections of information that happen to fit on one page. No one would call the phone book at topic, but it differs from these kinds of tables only in the number of entries it contains.

The use of paper tables for reference purposes is declining rapidly as connectivity and connected devices become more ubiquitous. You will rarely see an airline timetable, for instance. Rather, you will book your flight, including connections, using an interactive Web application. Train and bus timetables are also in decline as commuters increasingly rely on apps for directions and times. We need to start thinking of references not in terms of layouts on a page, but of interactions in an application.

We need to start thinking of references not in terms of layouts on a page, but of interactions in an application.

For example, compare the experience of looking for used cars in a paper magazine like Autotrader or the classified ads in your local newspaper with looking them up on an interactive website like AutoCatch. A paper source can organize cars by year or make or type or price or location, but it must select one axis. If you want to access the content along another axis – for instance, all the convertibles available in Ottawa with a manual transmission – the best hope you have is that the catalog provides an index along that axis. And even if it does, you have no way to collect all the entries together, so you end up with a catalog bristling with bookmarks or with your fingers awkwardly stuck in several pages.

And no paper catalog can do what Amazon does, which is to correlate its catalog with your previous purchases and the purchases of people whose buying habits overlap yours and then create a customized list of items that may be of interest to you.

We know that how content is organized and presented plays a big role in how easily readers can access, consume, understand, and act on the content. But users have different needs at different times and different users have different needs altogether. Therefore, no single arrangement of content can be optimal for all people at all times.

Now that we create content electronically and deliver it online, there is no reason to be bound by the limits of paper organization.

A database has the capacity to organize and present content to suit the present moment for the present reader. Paper can only present one generalized, compromise organization for everyone. Now that we create content electronically and deliver it online, there is no reason to bind ourselves to the limits of paper organization.

The term *database* is often associated specifically with relational databases, but this is only one type of database. A database is any collection of data that is structured so it can be queried reliably. While you could certainly use a relational database to structure and store your reference data, a suitably structured XML file or set of files can also be a database.

Creating reference content as a database will greatly enhance your options for display and navigation. As noted in Chapter 4, an API reference is a good candidate for faceted navigation, which requires a database on the back end. While a printed API reference is typically organized by library and by routine, readers may find other organizations equally useful. For example, readers may want to have a list of all routines that take or return a particular data structure. Listings like this are so useful that programming guides often include such lists in the text.

But in a doc set written in books, and equally in one written in topics, assembling such lists generally has to be done by hand, and the lists have to be maintained by hand when things change. If the API reference data were maintained as a database, those lists could be assembled on the fly by the doc build scripts.

Given the importance of the web for technical communications today, there is no reason to keep creating reference content in a form designed only for print-style publishing. It should be created in a form that supports interactive access. Even if you are not planning to provide such access today, you will surely be asked to provide it in the future, so you might as well start getting ready now. And, as a side benefit, you will find that structuring reference content as a database provides immediate benefits for your publishing process, particularly in the areas of accuracy, consistency, timeliness, and reuse.

To be absolutely clear, my point is not that a database is a nifty way to create a reference. The purpose of a reference is to allow people to look stuff up, and a database gives you more options than a paper layout for looking stuff up. A reference is a database by nature. A paper layout of reference information is an adaptation of the database to the limits of paper, both as a method of composition and a method of presentation. But we are no longer bound by the limits of paper. A reference is not a topic; it's a database. It's time to start treating it that way.

A reference is not a topic; it's a database. It's time to start treating it that way.

Everything else is not a concept

In any system that attempts to classify the whole of something, there is usually a category that essentially constitutes "everything else." In the trio of task, concept, and reference, that role belongs to concept. In Tom Johnson's shapes of help graphic (Figure 9.2), task has the shape of a procedure, reference the shape of a table, and concept the shape of plain text. Concept, then, stands for the plain, the generic, the feature-less and the structure-less.

But is that all there is to structured writing: tasks, references, and everything else? To be sure, for several reasons, every structured writing system needs some kind of generic topic structure:

- You will sometimes have topics that are narratives with no particular structure.
- You will sometimes have topics that could be given a structure, but you don't have enough of them to justify creating a separate type.
- You will sometimes have topics for which you plan to develop a structure, but you don't have enough information to develop that structure yet. A generic topic gives you a place to hold these topics until you can develop a structure.

Okay, but does it matter what you call that class of topics? Is there any difference between the terms *generic* and *concept*? Yes, it makes a difference in several ways.

First, the term *generic* reminds you that the topic has no type. The term *concept* makes it sound like the topic has a type, and that can lead you to think your topic-typing job is done.

You don't want to use a category name that only works because of its ambiguity.

Second, *concept* is one of the slipperiest words in the English language. Generally, it means a thought or an idea, but, as the Wikipedia article on concept[9] succinctly begins: "The word **concept** is defined variously by different sources." This multiplicity of meanings is perhaps what made it a candidate for the third member of the trio, since there is probably some meaning of the term that can be applied to any text. But you don't want to use a category name that only works because it is ambiguous. You don't want some members of the category qualifying under one meaning of the term and some under another. Categorization only tells you something useful when the categories are clear and unambiguous.

Third, one of the meanings of *concept* is relevant to many technology products. Many products are designed around certain fundamental ideas that users need to understand to use the product. When users switch from one technology to another – for example from an unstructured desktop publishing process to a structured writing process or from writing books to writing Every Page is Page One topics – they need to learn the fundamental concepts of the new technology and understand how those concepts differ from their old technology.

In this case, a concept is not "everything else" but something quite specific. Mixing "everything else" topics, even if they are useful, with these fundamental, "big-C" concepts is dangerous because the big-C concepts can get lost in the noise of hundreds of topics labeled *concept*.

Fourth, there are plenty of topic types that are not, by any reasonable definition, either tasks or references, nor are they big-C concepts. And they aren't generic either. An example from my background writing about programming languages and operating systems is the annotated code-sample topic.

Programmers want two things above all else in a documentation set: a thorough API reference and working code samples. Programming task topics may be useful, and in

9 http://en.wikipedia.org/wiki/Concept

some cases, programmers would benefit from understanding the fundamental concepts of the language or OS. However, what every programmer wants is an API reference and code samples.

Code samples deserve a topic type of their own. They have a definite structure that includes the sample code itself, including annotations, plus information about supported versions, programming language, resources required, and performance and security characteristics.

The pathfinder topics discussed in Chapter 15 also deserve their own topic type. Pathfinder topics are not task topics (in particular, they should not be confused with workflow topics, which are a kind of task topic), nor are they big-C concept topics. Rather, they are topics that help users find their way around the technology, giving them a sense of where everything is and how all the parts work together. If big-C concept topics are aimed at understanding, pathfinder topics are aimed at orientation and planning.

If big-C concept topics are aimed at understanding, pathfinder topics are aimed at orientation and planning.

There are many topic types that don't fall into the task or reference categories, but only one qualifies as a true *concept*. Concept is not the right word to describe "everything else." And it isn't useful to single out tasks and references as particular types and then assign everything else to a single category, however you name that category. To do so is akin to dividing the animal kingdom into cats, dogs, and everything else.

CHAPTER 10

EPPO Topics Establish their Context

Because readers may come from anywhere, and often arrive at a topic through an imprecise mechanism such as a Google search, a topic should clearly establish its context in the subject domain. As noted in Chapter 6, the scent of information is key to the information forager finding your content. Properly establishing the context of your topic in the real world is a key part of making it smell right.

If you have ever landed in the middle of a help system from a search and found that you have no idea where you are, you have experienced the lack of context that so many topics exhibit.

For example, Figure 10.1 is a topic from the Eclipse help system. This topic provides little context and clearly isn't an Every Page is Page One topic. Unfortunately, topics like this are common in help systems.

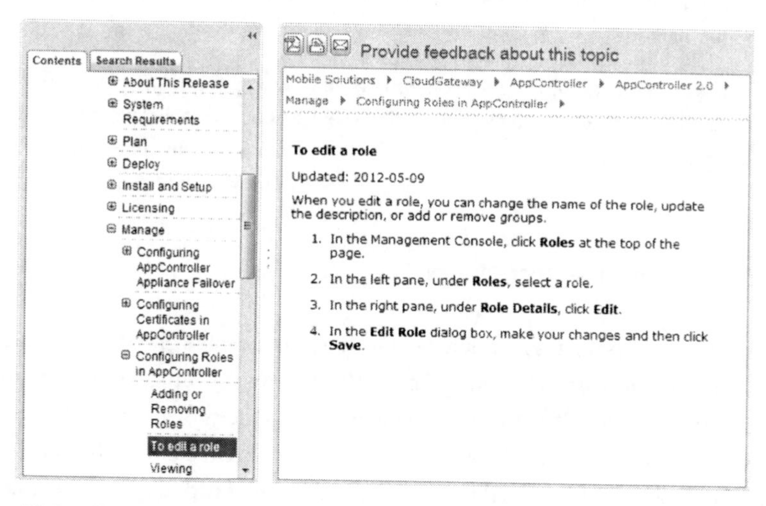

Figure 10.1 – Edit a role topic

Establishing context

A self-contained topic must establish its context, and readers must be able to come to it from anywhere and know where they have arrived. Most Every Page is Page One topics orient themselves quickly. A lead paragraph of a sentence or two often suffices to set the scene for what is to come.

Figure 10.2 shows a good context-setting example from the Google App Engine documentation. Here the context is the class of data storage problem the topic addresses. It tells the reader what class of problem is going to be discussed, and in what environment that problem can occur. Without this introduction, the reader of the topic might be confused about what part of the problem space the topic is addressing.

> Storing data in a scalable web application can be tricky. A user could be interacting with any of dozens of web servers at a given time, and the user's next request could go to a different web server than the one that handled the previous request. A web server may depend on data that is spread out across dozens of machines, possibly in different locations around the world.
>
> Thanks to Google App Engine, you don't have to worry about any of that. App Engine's infrastructure takes care of all of the distribution, replication and load balancing of data behind a simple API—and you get a powerful query engine...
>
> —Google, *Using the Datastore*[1]

Figure 10.2 – Context-setting example

Putting a topic in context does not mean placing it in the table of contents or locating it in the information set in any way. Figure 10.1 locates the topic within the table of contents, but this does little to tell you where it fits in the subject matter. A reader arrives at your topic looking for information, not a place in a book. Ultimately, readers are looking for knowledge about the real world. Putting a topic in context means locating the subject of the topic in the real world. Placing an **Up to TOC** link on a topic does not place that topic in context.

[1] https://developers.google.com/appengine/docs/go/gettingstarted/usingdatastore

There are many mechanisms you can use to establish context. A good title is a great start. A succinct, context-setting first paragraph, as in Figure 10.2, is also important. Good context setting clearly and obviously differentiates Every Page is Page One topics from topics that were created by bursting books. Figure 10.3 is an example of context setting from the Wikipedia article on Ottawa.

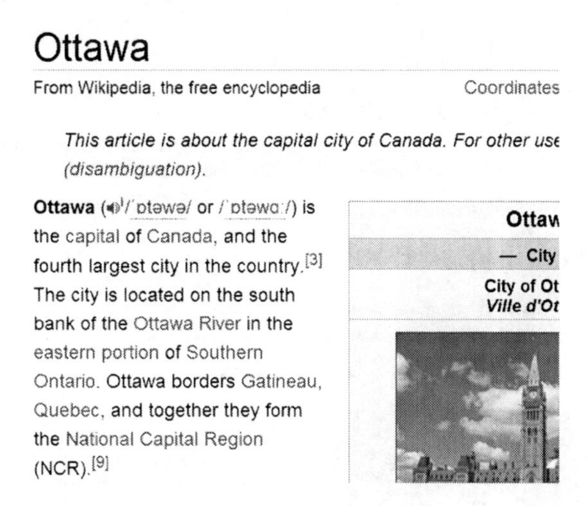
Ottawa
From Wikipedia, the free encyclopedia
Coordinates
This article is about the capital city of Canada. For other use (disambiguation).
Ottawa (/ˈɒtəwə/ or /ˈɒtəwɑː/) is the capital of Canada, and the fourth largest city in the country.[3] The city is located on the south bank of the Ottawa River in the eastern portion of Southern Ontario. Ottawa borders Gatineau, Quebec, and together they form the National Capital Region (NCR).[9]
Ottaw
— City
City of Ot
Ville d'Ot

Figure 10.3 – Wikipedia article on Ottawa

The first paragraph in Figure 10.3 locates the city of Ottawa geographically and politically. You know at once which Ottawa you are dealing with. In a couple of brief and brisk sentences, the reader knows exactly where this topic fits.

Another way to establish context is to use a graphic. The context of a recipe is usually clear, but you can make it even more clear and easy to digest with a picture of the dish, as in Example 7.1. One glance at the picture and you know if it is a dish you want to make.

Metadata[2] is yet another mechanism for establishing context. A good example of this the entry for the Blue-Footed Booby[3] from *All About Birds* (see Figure 10.4). The

[2] Metadata consists of labels attached to content to identify the subject matter – more on metadata in Chapter 19.

[3] http://www.allaboutbirds.org/guide/Blue-footed_Booby/id

place of the Blue-Footed Booby in the Linnaean taxonomy of animals is shown as part of the frame around the content. Using a metadata frame makes the information available in a consistent place on each page, without taking space away from the body of the topic. This example also makes the context navigable by providing links to browse the taxonomy by name and shape and to view similar and related species.

Figure 10.4 – Blue-Footed Booby

As with faceted navigation, metadata only works to establish context when readers are already familiar with the taxonomy on which the metadata is based. If the taxonomy is meaningless to your readers, then it won't provide any useful context. However, you can provide multiple taxonomies to serve different readers. For example, *All About Birds* lets you establish context with the Linnaean taxonomy, if you know it, but it also gives you a picture and the opportunity to use alternative metadata such as shape and color.

Context and the imprecision of search

When you find documentation on the Web, search will sometimes land you in the documentation for a different version of a product than the version you own. There are two problems here. First, in burst-book content, an individual page may not identify which version of the product it applies to. Secondly, if you land on a page for the wrong version, there may be no convenient way to get to the equivalent page for the right version.

If you land on the page for a not-quite-right bird in *All About Birds*, you have lots of tools for getting to the right bird. If you land on a not-quite-right topic in most online help systems, you often have no way to get to the right topic.

Atlassian handles this very well in the Confluence documentation. If you hit a page that is not for the current version, you get a banner at the top of the page advising you of this and a link to the same page in the current documentation (Figure 10.5). The only improvement I can suggest would be to offer a drop down so you can go to the same page for any version.

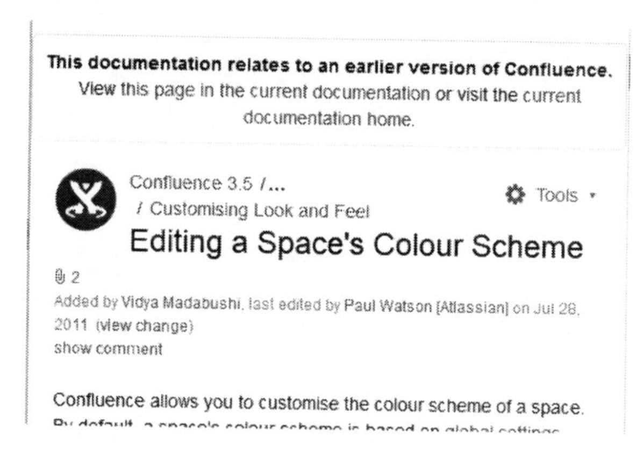

Figure 10.5 – Context-setting link to current version

The Wikipedia article about Ottawa (Figure 10.3) has another nice context-setting feature that I think should be imitated in every large content set. If a word has more than one meaning within the content set (in this case, if there is more than one item in the encyclopedia with the name Ottawa), then there is a disambiguation line above the content that states which article this one is and offers a link to a list of the other articles on this subject.

> This article is about the capital city of Canada. For other uses, see Ottawa (disambiguation).

This highlights a problem Google and other search engines have. They always return the most popular results. After all, that's what they are designed to do. However, this means that less common subjects that happen to share terminology with more common subjects get pushed far down in the search results. You need significant search skills to compose a search string to get the results you need. But with Wikipedia, you don't need those skills, because you can browse the more obscure topics right at the top of the most popular topic. This is the sort of thing we should all be doing.

CHAPTER 11

EPPO Topics Assume the Reader is Qualified

Authors tend to write books assuming they will be read straight through by readers with a wide variety of backgrounds and skills. Therefore, authors often assume their typical reader is not fully qualified to read the book, and they spend a great deal of time and text explaining basics and background material. Of course, this annoys and delays the more qualified reader. That is the peril of delivering a single curriculum for all readers.

This approach is not appropriate for Every Page is Page One topics, and it violates many of the other properties of EPPO topics, such as having a specific and limited purpose, staying on one level, and conforming to a type. An EPPO topic should be written for a qualified reader.

Example 7.1, "Tarragon Mac and Cheese Recipe," contains the instruction "Cook your pasta." It does not tell you how to cook pasta. It assumes you know or can figure it out for yourself. It makes similar assumptions with the following instructions:

- "Whisk in the flour ..."
- "Simmer for 10 minutes ..."
- "Crack the egg ..."
- "Add the onions and sweat ..."

Know how to whisk? Know what simmer means? Know how to crack an egg (without getting the egg on the floor or the shell in the pan)? Know how to sweat an onion? This recipe is written for an experienced cook, and it assumes that an experienced cook knows how to do all these things.

If you are not as experienced as the topic assumes you are, are you out of luck and unable to cook yourself some Tarragon Mac and Cheese? Of course not. You can look up all of these things and learn how to do them. If you need to find out how to sweat an onion, search for "sweat an onion" and you will find an excellent Every Page is Page One topic from the Reluctant Gourmet[1] (shown in Figure 11.1).

Figure 11.1 – How to sweat vegetables

Having learned what you need to know, you can then go back to the Tarragon Mac and Cheese recipe and prepare it successfully.

---

[1] http://reluctantgourmet.com/cooking-techniques/more-specific-techniques/item/60-how-to-sweat-vegetables

Of course, when it comes to technical content, the Web does not always provide an answer. Some things may be unique to your product, in which case your doc set needs to provide them. But the same principle applies. Individual topics should make appropriate assumptions about the qualifications of the reader. If there is a possibility that not all your readers will be qualified, you should provide the topics they will need to qualify themselves. Then you should make sure readers can find them.

When you write prerequisite topics for users who are not qualified to read your initial topic, those topics should also be Every Page is Page One topics, and they should assume that their readers are qualified to read them. And if you need another set of topics to prepare the readers to read these topics, create them as well.

The How to Sweat Vegetables topic serves a reader's *derived purpose*. Sweating vegetables will never be a reader's main purpose. However, when a reader needs to sweat vegetables for any recipe, this topic can serve that derived purpose. Serving readers' derived purposes is a major part of what technical communicators should be doing.

Reader dependencies vs. subject dependencies

When dealing with qualification, you need to consider the difference between a topic having a dependency and a reader having a dependency. Fair readers will blame a topic for not fully covering its subject. They will not blame the topic for their own dependencies.[2]

The need to sweat onions before adding them to the Tarragon Mac and Cheese recipe is a subject dependency. The dish won't taste right if this is not done. It is knowledge specific to preparing this dish, not something a cook would infer naturally. No matter how experienced a cook you are, the recipe still has to tell you to do this.

Not knowing how to sweat an onion is a reader dependency. It is part of the general knowledge of cooks, and if you don't have it, that is something missing from your

[2] Though we should note that the less qualified readers are, the less able they are to judge what is a reader dependency and what is a subject dependency.

cooking knowledge. Within the community of cooks, this knowledge deficit is specific to you, not to any recipe that calls for the sweating of onions or other vegetables.

Often, the feedback technical writers receive from customers – either directly or through sales or support – concerns reader dependencies. When you get this kind of feedback, you may be tempted to add some information to the manual where the reader got stuck. With paper manuals, this may be justified. It can take considerable effort for a reader to find additional information outside the manual.

The problem is, every time we stick these bits of information into the manual, it grows in both size and complexity. And because you hear about some reader dependencies and not others, your manual becomes increasingly uneven. Within a single procedure, you may end up with one step described in excruciating detail and the next treated at a very high level. You might have one feature described for a grade school audience and the next for a grad school audience.

Imagine what the Tarragon Mac and Cheese recipe would look like if all the potential reader dependencies were included. It would be many pages long, and it could be difficult for an experienced cook – for whom the recipe was written – to find the actual preparation steps among all the asides and additions. An EPPO topic has to assume the reader is qualified and refrain from trying to meet reader dependencies for readers who are not. Otherwise the topic will cease to be manageable for qualified readers.

When we say a topic is self-contained, we mean that it is free of subject dependencies. We do not mean, and cannot reasonably demand, that it is free of reader dependencies. Most readers will have dependencies. To meet those dependencies, the reader is sometimes going to need to consult other topics. To assist them, a good EPPO topic links richly to ancillary topics, something I will discuss in Chapter 13.

To be considered self-contained, a topic must meet the reader's reasonable expectation of a topic of this sort. It does not have to satisfy all the reader's personal dependencies.

Determining the qualified reader

Determining the qualified reader is not an arbitrary or subjective process. It follows from the specific and limited purpose of the topic. A qualified reader is a reader who knows everything needed to perform the specific and limited purpose of the topic except the specifics of the case that the topic covers.

The appropriate level of qualification for a topic is probably best assessed at the level of someone who does this task regularly. Of course, you need to define what "regular basis" means for each task. Some tasks are done more frequently than others. What you are looking for is the level of qualification for someone who does this task on its natural schedule. If the natural schedule is once a day, the level of qualification is likely to be much higher than if the natural schedule is once every five years.

Another factor is the typical occupation of the person who does this task. For instance, someone who does a task once every three months might not retain much knowledge, but someone who does similar tasks with similar objectives and similar tools every day – and those tasks rely on similar knowledge – will probably retain much more.

Choosing the level of understanding

Of course, the same idea can be explained at different levels of understanding. For example, consider the article "Why You Generally Shouldn't Put Metals in the Microwave"[3] from the website *Today I Found Out*. The article clearly assumes some background in physics and electronics:

> First, let's talk a little about how a microwave oven actually works. At its core, a microwave oven is a pretty simple device. It's basically just a magnetron hooked up to a high voltage source. This magnetron directs microwaves into a metal box. These generated microwaves then bounce around inside the microwave until they are absorbed via dielectric loss in various molecules resulting in the molecules heating up.

[3] http://todayifoundout.com/index.php/2010/08/why-you-generally-shouldnt-put-metals-in-the-microwave

Since it assumes we know what "dielectric loss" means, the article is clearly aimed at a geekier audience than the average cook.

Obviously, the article could be written for a less qualified audience. However, in the process it would either get a lot shorter, by omitting the technical details, or get a lot longer, by attempting to explain what things like dielectric loss mean. In other words, you can't just translate this article into simpler language and expect it to do the same job for a wider audience.

The article does not just explain why you should not put metal in your microwave, it explains the mechanisms in enough detail that a suitably qualified reader should be able to extrapolate from this information to understand related concepts or to figure out when it might be okay to put metal in the microwave. Change the audience and you don't just change the vocabulary, you change the level of interest and the level of ability to extrapolate from the explanation, which is, in turn, a change in the purpose of the topic.

Therefore, when you select an audience for your topic you are not just selecting the vocabulary, you're also making assumptions about the level of interest and the degree of extrapolation that audience is capable of. If you are writing a general interest topic on the Web, that decision may be more or less arbitrary. But if you are writing for a technical audience, these assumptions aren't arbitrary, they are directly related to the task. The task tells you the level of interest and the level of extrapolation expected, which in turn tells you who normally does this job, what they know, and what they expect to be able to do with the information you give them.

Unless you can address each individual, you have to write for an aggregate level of interest.

Of course, individual users differ greatly in their levels of technical knowledge and interest. Unless you can address each individual, you have to write for an aggregate level of interest. Generally speaking, you want to aim for a level of knowledge that will enable the reader to accomplish a new task or attain a higher level of productivity with an existing task. In the end, you should enable action and not accommodate infinite varieties of curiosity. Focus on the levels of knowledge required to accomplish concrete tasks.

Avoid arbitrary labels

In the same vein, the labels *novice, intermediate,* and *expert* are of no help in determining qualification level. These are judgments, not qualifications. In *The Nurnberg Funnel,* John Carroll makes the following argument:

> The term *novice* is problematic for designers of training. Its use exposes a technocratic ideology of learning that is insulting. Adult learners can never be thought of as novices. They are experts, though perhaps in domains other than the one in which they are training.
>
> —John Carroll, *The Nurnberg Funnel*[8, p. 87]

People don't take on a task based on the documentation writer's idea of whether it is a novice, intermediate, or expert level task. A person's responsibilities and ambitions are not calibrated to the capabilities of your product or to the organization of your documentation set.

The qualifications for a task are specific to that task. Some qualifications are easy to obtain, while others are more difficult. And each reader will come to the topic with a different set of qualifications. A good EPPO topic set allows each reader to choose a unique path based on his or her information needs. Artificially defined levels like novice, intermediate, and expert don't help readers create a unique path.

If your workplace has well-defined roles with separate responsibilities, then it can make sense to use those roles to define the qualified user for each topic or topic type. But artificial or poorly defined classifications will only confuse and frustrate users.

Qualification and findability

We know that many searchers are not completely qualified for the task they want to accomplish. They often lack some background information or skills. Even so, findability is improved when you assume that searchers are well qualified. Why is this the case? Quite simply, because searchers assume they are qualified. If they didn't, they would be searching for something else.[4]

The way you reach your readers is to write topics on subjects they are interested in and assume your readers are qualified. That is the only way you will capture their attention. Then, provide a clear context statement in each topic. This should enough to signal unqualified readers that they need more background. If you then provide links to prerequisite material, your readers can get to the information they need to become qualified.

[4] If you know you are unqualified for a particular task, you will probably search for information that will help you become qualified, and you assume you are qualified to do that search. So, you nearly always assume you are qualified for what you are currently searching for.

CHAPTER 12

EPPO Topics Stay on One Level

There are multiple levels to every subject: levels of detail; levels of abstraction; strategic, tactical, and operational levels of interest; even subject matter of interest to different levels of an organization or pertaining to different layers of a multi-layered systems or the different roles that operate on those layers. Most people need information on more than one level in order to complete all of their tasks, or a single complex task. However, it is preferable for a topic to stick to one level.

Changes of level are a necessary part of any course of study. As you study a subject, you sometimes need to dive down into the details in order to get a practical illustration of a general principle. When you are working on some detail, you sometimes need to understand a more general principle that explains why the detail works the way it does (and not the way you expected it to).

If you are hungry for pizza, you are not interested in meeting the cow that gave the milk for the cheese or the farmer who grew the tomatoes.

But if you are hungry for pizza, you are not interested in meeting the cow that gave the milk for the cheese or the farmer who grew the tomatoes. People who search for information have already selected the level of abstraction they are interested in by how they frame their query or how they select among search results. If your topic starts off looking like pizza but changes to cow after two paragraphs, readers will not be pleased. If they wanted cow, they would have searched for cow.

Each reader will decide when to change levels, depending on personality and circumstances. Some readers prefer to absorb the big picture before acting. Some prefer to be immersed in the details and integrate their understanding of the whole at a later time. Some only want to follow procedures without understanding. Some are in a hurry to complete an individual task. Some are preparing for a project and want a broad view so they can plan the best strategy overall. The decision about when to change levels, therefore, is best left to the individual.

Example 7.1, "Tarragon Mac and Cheese Recipe," contains the instruction "Cook your pasta." It does not dive down into the details of how to cook pasta. That is a more basic recipe, one probably best obtained from the box the pasta came in. It would make no sense for the Tarragon Mac and Cheese recipe to go into the details of cooking the pasta. The recipe wisely stays on the higher level of covering the preparation details that are particular to Tarragon Mac and Cheese.

Similarly, the WordPress Codex article on Using Themes (Example 8.1) stays on one level. For instance, in the introduction it presents users with a basic overview of themes, which is sufficient for them to decide if they want to install themes.

> A WordPress Theme is a collection of files that work together to produce a graphical interface with an underlying unifying design for a weblog. These files are called **template files**. A Theme modifies the way the site is displayed, without modifying the underlying software.
>
> —From Example 8.1, "Outline of Using Themes for WordPress Codex"

If this article went into the details of how a template file is constructed, that would change levels. You don't need to understand the details of template files to install and use themes. But the article does provide a link to the topic on template files, so a reader who wants those details can move to that level.

Books change levels at the author's fiat

Most books don't stay on one level.[1] In books, this one included, it is the author who decides when to present the big picture and when to delve into details. This book begins at a high level, talking about how accessing and using information is different in the context of the Web and now is down in the details of the characteristics of individual topics. Later it will expand its view to discuss how topics work together and how you go about writing and managing topics. Planning when to make these changes of level was a big part of the planning and writing effort for the book and the subject of much discussion between my reviewers, the publisher, and me. I decided, in advance, when to show you the big picture and when to turn your attention to the details.

---

[1] Cookbooks do, because cookbooks are really just collections of topics, not sustained narratives.

This is a necessary consequence of the linear structure of a book. Most books are designed to be read in a particular order, and that means the book will change levels when and where the author chooses.

Of course, the writer is sometimes in a better position to make these decisions than the reader. The writer has a more holistic view of the material, can anticipate particular points that might be confusing, and can choose to present ideas in a sequence that avoids confusion. This is one of the hallmarks of a great book. On the other hand, few authors have that kind of talent, and impatient readers, trying to get their work done, seldom have the patience or subservience to subject themselves to the author's curriculum, however well planned it may be. And today they are more free than ever to take command of the curriculum for themselves.

The tendency to change levels seems to be true of most books, but it is true of technical manuals particularly. The typical organization of a technical manual is to break down the subject into functional areas and devote one chapter to each. Each chapter begins with the introduction of that functional area and then drills down into finer and finer levels of detail.

Here is an example drawn from a book pulled more or less at random from the programming section of my bookshelf. Example 12.1 shows a section of the table of contents from Bruce Eckel's *Thinking in Java*:

Example 12.1 – Technical manual TOC example

7: Polymorphism
- Constructors and polymorphism
  - Order of constructor calls
  - Inheritance and finalize()
  - Behavior of polymorphic methods inside constructors
- Designing with inheritance
  - Pure inheritance vs. extension
  - Downcasting and run-time type identification

Here we see a clear change of levels. He starts with the broad concept of polymorphism, drops levels to a more specific discussion of the relationship of constructors and polymorphism, and then drops down again to a detailed consideration of one function. Then the chapter pops up a level to talk more abstractly about design issues and then drops down again to a discussion of downcasting and run-time type identification.

This level changing is typical of how a chapter in a book is constructed. But it is completely different from how an independent topic is constructed. For instance, see the Wikipedia article on polymorphism,[2] which stays on the level of describing the general concept. It does include pseudo-code examples (Example 12.2) to illustrate what it is saying, but never delves down into the details of syntax.

Example 12.2 – Polymorphism code example from Wikipedia

```
program Adhoc ;

function Add ( x , y  : Integer ) : Integer ;
begin
    Add  : = x  + y
end;

function Add ( s , t  : String ) : String ;
begin
    Add  : = Concat ( s , t  )
end;

begin
    Writeln (Add ( 1 , 2 ) ) ;
    Writeln (Add ( 'Hello, ' , 'World!' ) ) ;
end.
```

The article is too long to include here, and, because Wikipedia articles are constantly changing, it may no longer be on one level when you read it. One of the interesting things about the Wikipedia process is that an article may not meet all the EPPO criteria all the time, but, through the refactoring process, most articles take on EPPO characteristics with remarkable consistency.

[2] http://en.wikipedia.org/wiki/Polymorphism_(computer_science)

Writing on one level is not an artificial discipline. It is what writers naturally do when they view their work as an independent topic. The challenge for technical communicators, especially those who have spent a career writing books, is to think of their project as a set of independent topics. We will look at tackling a large subject with many independent topics in Chapter 15.[3]

Keeping topics on one level

Keeping a topic on one level can be a particular challenge for writers used to creating books. The key to creating topics that stay on one level is to keep in mind the following characteristics of Every Page is Page One topics, which should be familiar by now.

- **Self-contained:** In Chapter 7, we saw that an Every Page is Page One topic is like a car, not a brake caliper; it is a unit of content that can function alone. By the same token, when you are on a journey, you may need both a place to sleep and a way to move about. But you do not add beds and bathrooms to a car. Do that and the car becomes an RV, a much more expensive and ungainly vehicle that is unsuited to the other things you use your car to do, like running to the store or picking up the kids from school. An RV is not a topic. An RV is a honking big book. It is a vacation reading project, not quick help. As a self-contained transportation device, a car sticks to the transportation level and leaves the accommodation level to others. If your topic starts to feel like a car with a mattress strapped to the roof, that is a good sign that you are changing levels, and it's time to create a new topic.
- **Specific and limited purpose:** There is a very good reason that the title of Chapter 8, contains two separate adjectives: *specific* and *limited.* No matter how specific your purpose may be, there is always a way to justify adding ancillary and supporting material. The word *limited* is vital, and the limits it imposes may be seen as chiefly vertical. Yes, there are always larger questions behind the specific purpose of a topic, but the limits exclude them. Yes, there are always specific details, local exceptions, and interesting tidbits of information, but the limits exclude

[3] Note how this forward reference acknowledges that we have just touched on a higher-level concept and how it puts the reader off from pursuing it now by promising to treat it later.

them. Sometimes it is necessary to discipline yourself by writing down the limits you put on each topic type. "Don't put the biological taxonomy of the ingredients in the recipe," "Don't include the entire command syntax of every tool used in this task," or "Don't add anything just because the product manager said it ought to be in the manual somewhere."

- **Conform to type:** The type of a topic, as described in Chapter 9, is one of the best ways to keep your topic on one level. If some material you are trying to add seems to strain the boundaries of the type, there is a good chance you are trying to change levels. One reason to use structured writing techniques and define your topic type as specifically as possible is that you can more easily see when you are going beyond the bounds of the topic type. On the other hand, if you find your topic starting to change levels, and the topic type is not pushing back, you probably need to tighten your topic type definition to keep other topics from straying away from their proper level. For more on structured writing, see Chapter 18.
- **Establish context:** Establishing the context of a topic (Chapter 10) also helps keep it on one level. The context of the topic orients readers, giving them a sense of whether or not they are qualified. By orienting your readers, you also orient yourself and set your expectations about what this topic needs to include (or exclude) to meet its specific and limited purpose. In addition to orienting readers, the context setting part of your topic is the best place to provide links that can help unqualified readers find the information they need to become qualified. By getting these issues out of the way at the beginning, you can assuage your own fears about readers not being able to follow the rest of the topic.
- **Assume qualified:** In Chapter 11, we saw that an EPPO topic assumes that the reader is appropriately qualified to complete the specific and limited purpose. When books or topics change levels, it is usually because the author was worried that some readers wouldn't know how to sweat an onion. In an Every Page is Page One topic collection, the correct way to accommodate unqualified readers is to provide topics readers can use to qualify themselves and links to those topics (bottom-up navigation). The temptation to change levels in a topic almost always springs from the fear that a reader may not be qualified to understand something you have just written. In those moments, it is important to have an outlet for that fear. If you are not sure that all readers will be qualified to understand a point, record it in a list of prospective topics (preferably a centrally maintained one).

> Capturing the possible qualification deficit is important because it is difficult to anticipate all the qualification deficits readers may face. When you discover one, always write it down and share it with the rest of your team. This will help you build a picture of who your readers are and what they need. But, equally important, don't handle the deficit yourself by changing levels in your current topic. Not only does that distort the topic, it means that the information is lost to the rest of the team. A topic that might serve many purposes besides supporting your topic may never get created.

This is not to say that changing levels is wrong when you design a book or a chapter. In many respects changing levels works well when a book is read in the order the author intended (as I am sure you are diligently reading this book). However, it doesn't work in EPPO topics. Topics and chapters are fundamentally different beasts, and you cannot make good topics by chopping up books.

CHAPTER 13

EPPO Topics Link Richly

> Links are the visible manifestation of the author giving up any claim to completeness or even sufficiency; links invite the reader to browse the network in which the work is enmeshed, an acknowledgement that thinking is something that we do together .
>
> —David Weinberger, *Too big to know*[27]

Linking is surprisingly controversial in technical communication and content strategy. Many writers feel that links act simply as a distraction, tempting the reader to leave the current text.[1] Information foraging would suggest that the reason people leave an information patch is not the presence of links, but the lack of tasty information that is easy to catch and easy to digest. To put it more directly, people get distracted when content is boring or useless.

However, information foraging theory also tells us that the easier it is to move to a new information patch, the more likely it is that the reader will abandon the current patch for another. So it is probably true that linking will lead people away from the content they are reading if that content is not very nutritious. The question is, should we care?

If you struggle long and hard to create something, you will naturally be offended when people don't seem to appreciate your work. If you have spent hours in the kitchen preparing an elaborate meal, only to have your family pick at it and then complain half an hour later that they are still hungry, you have felt the pain of under-appreciated labor. That pain has probably led you to tell your family to shut up and eat. It's an understandable feeling.

---

[1] There is a good deal of research on the effects of linking in Web content, and not all of it points in the same direction. I surveyed the literature and provided a critique in an article on The Content Wrangler, "Re-Thinking In-Line Linking: DITA Devotees Take Note!" [http://thecontentwrangler.com/2012/05/03/-re-thinking-in-line-linking-dita-devotees-take-note/]

We have always known, of course, that users seldom sit down and read our manuals as they were designed to be read. They pick at them here and there, ask for more examples, and then complain half an hour later that they can't get the product to work. "Read the fine manual," we mutter under our breath. This may be how users always have behaved, but that does not stop us from resenting them, nor from trying to design our information products to prevent them from doing it.

Information foraging is the optimal information seeking behavior for most people most of the time.

An Every Page is Page One information design, on the other hand, does not start from the premise that the goal is to stop the user from wandering. Rather, it starts by acknowledging that this is how users behave and recognizing that, based on the limits of knowledge and the urgency of their tasks, information foraging really is the optimal information seeking behavior for most people most of the time. However, even if you don't agree that information foraging is optimal, it is still how users behave. We can't beat it. We've tried every trick in the book to no avail. It's time that we started to facilitate it instead.

Every Page is Page One information design is built around two propositions: 1) the way to keep readers is to provide the content they need, and 2) if readers want to move to content that better meets their needs, we should help them get there. That means linking richly.

Linking is not an incidental feature of EPPO topics, it is a key part of EPPO design.

Because an EPPO topic does not have parent or child topics, it relies on linking to locate itself in its subject area, to establish itself in its context, and to allow readers to navigate effectively. Linking is not an incidental feature of EPPO topics (as it often is in book-structured material), it is a key part of EPPO design. In a bottom-up information architecture, the primary means of navigation is from the current page, which serves as a hub of its local subject space.

From the author's point of view, links serve to keep readers in your content. A foraging reader is more likely to move to a new patch when it's easy to get there. By providing links, you can lead readers to other information patches you own, reducing the temptation for them to move to a competitor's content.

Links and the democratization of knowledge

It is also worth noting the difference between the paper world and the Web when it comes to the democratization of knowledge. In the paper world, it was accepted that certain knowledge essentially belonged to certain groups. To gain access to that knowledge, you were expected to follow a prescribed curriculum that qualified you for membership in those groups. It wasn't that the knowledge was physically locked away, it was that the knowledge was written in a way that assumed you had followed the curriculum. It was written for insiders.

For centuries, content tended to be published in specialized journals for specialized audiences whose ability to read them could be assumed with a reasonable degree of certainty. Even the knowledge of the existence of such journals, and the topics they covered, was largely confined to the academic community. And in any case, there was no recourse, because the cost of paper meant that no space could be spared to bring the unqualified reader up to speed. If you were not qualified to read the journal, it was up to you to go find yourself an education.

Content on the Web, however, is published to the whole world. Anyone, regardless of background or qualifications, can land on your content. You can, of course, hide content behind a login, but even then it is consumed in the context of the Web. By hiding content you don't restrict it to qualified people so much as hide it from everybody.

Anyone, regardless of background or qualifications, can land on your content.

So now you have a flood of unqualified readers finding your content. But now they are not shut out by their lack of qualification. The whole of the Web is available to them to find the information they need to qualify themselves to read your content.

Thus, the Web democratizes information both by making information easy to get and by making obscure information easier to decode and burrow your way into. Some may despair at this, preferring a world in which expertise is left to the experts and dilettantes are left in the cold. Much as been written on the virtues or vices of this development.[2] The debate is outside the scope of this book, but there is no question

[2] For the case for vices, see Nicholas Carr's *The Shallows*[7]. For the case for virtues, see David Weinberger's *Too Big to Know*[27].

that links are the great democratizing elements of the Web. (In David Weinberger's words, "hyperlinks subvert hierarchy."[16]) From a commercial point of view, that means that links make your content available to more people, and that's a good thing because it helps your company sell more stuff.

Linking and findability

In bottom-up navigation, which your readers are using whether you are helping them do so or not, links are an essential part of findability. When readers arrive at a topic, they may be close to the information they need, but they may not be in quite the right place or they may be hungry for more. In either case, they need a way to move easily to the richest information patch for their needs.

Jared Spool has found that readers are much more successful at finding content using links than using a local site search.

> Overall, users found the correct answer in 42% of the tests. When they used an on-site search engine (we did not study Internet search engines), their success rate was only 30%. In tasks where they used only links, however, users succeeded 53% of the time. ... our testing data suggests that designers would have more success by focusing instead on creating effective links.
>
> —"Why On-Site Searching Stinks"[3]

You still have to provide appetizing content at the end of your links, of course. Each link should point to a good Every Page is Page One topic that supports the reader's purpose for selecting it. Providing useful links to good content will make your readers' lives easier and reduce the temptation for them to leave your content.

[3] http://www.uie.com/articles/search_stinks/

Therefore, links play a larger role in EPPO content than cross references or footnotes play in books. You should be thinking of links not as citations or references but as the natural expression of every significant subject affinity in your content.

You should be thinking of links not as citations or references but as the natural expression of every subject affinity in your content.

In particular, there are two important reasons why your context-setting material should be rich with links:

First, readers are most likely to discover that they lack qualifications while they are reading your context-setting material. The context-setting section, precisely because it locates the topic in its subject area, is rich in subject affinities. If those subject affinities are linked, they will supply much of what your readers need to become qualified.

Second, search, and other methods of finding content, can be imprecise, because of both the limits of search engine technology and the limits of readers' skills in framing search terms. This imprecision may land readers on your topic when they really wanted a related topic. Linking to contextually near topics gives readers the means to travel the last mile to the content they really need.

Links are also important for those who have eaten but are hungry for more. Links along lines of subject affinity make it easy for them to find their next meal. Of course, you can't tell readers what they should do next. It is their curriculum, not yours. Link along lines of subject affinity and let readers choose which road to take.

A systematic approach to linking along lines of subject affinity is critical here. As Sean Carmichael notes, summing up a keynote presentation by Jared Spool:

> Websites are full of links. How useful these links are in helping users complete tasks is another story. Links have to guide users as they follow the scent of information. A vague or confusing link often leads users down a wrong path and in turn increases their rate of failure.
>
> —Sean Carmichael, "Jared Spool – The Secret Lives of Links"[4]

Links should help readers follow the scent of information and navigate the lines of subject affinity between topics.

---

[4] http://www.uie.com/brainsparks/2012/12/12/jared-spool-the-secret-lives-of-links/

Writing Every Page is Page One Topics

How do I go about writing Every Page is Page One topics, and how do I cover a large subject with only topics?

CHAPTER 14

Writing Every Page is Page One Topics

The Every Page is Page One topic is not a new invention. As the examples in the previous chapters have demonstrated, Every Page is Page One topics can be found all over the Web. Nor are EPPO topics unique to the Web. Essays and articles in journals and magazine have followed the Every Page is Page One format for centuries. The difference today is the ease with which readers working in the context of the Web can move around from one piece of content to another. This behavior puts a premium on content that supports this style of reading – that is, Every Page is Page One content. Whether you are delivering your technical content on the Web today, tomorrow, or never plan to, your readers are reading in the context of the Web, and you will serve them best by writing EPPO topics.

There is nothing new about writing EPPO topics. It is something most skilled writers do naturally if they are writing a single article or contributing an entry to Wikipedia. What writers often find more difficult is to create a set of EPPO topics to cover a broad piece of subject matter such as the documentation for a major product.

Textbooks vs. user assistance

Most traditional user manuals follow what we might call the textbook model. Implicitly or explicitly they are designed to be read in a specific order. Even though writers may not expect someone to read the manual sequentially, that is the model of composition they use. Textbooks are top-down and isolated from other content.

Many discussions of the advantages and disadvantages of topic-based writing seem to neglect the different view of the user that is inherent in the move from the traditional textbook style manual to stand-alone topics, and in the move from top-down media like paper and help to the bottom-up character of the Web and web-like systems. Topics are not simply a new mechanism for composing and constructing documents, nor are they simply about enabling reuse. Nor are they even about moving content to the Web. What the move to topics is really about is a move away from the textbook model of documentation towards a user assistance model.

The textbook model assumes readers want to learn about a subject, and that if they are going to act on what they have learned, they will do so afterwards. The user assistance model assumes readers are working, have hit a snag, and need immediate aid. It assumes that readers will plunge into the work, as far as working conditions allow them to, and use any resource they can find to get moving forward again.

Of course, this does not mean that those who write textbook-style manuals actually expect, or are even trying to foster, learn-first-do-afterwards behavior. Writers who are students of minimalism will no doubt encourage exploration and will want to help people troubleshoot when they get stuck. However, this is difficult within the confines of the top-down textbook form of a traditional user guide.

There has been a user assistance component to documentation for a long time, and good user assistance has always been topic based.[1] But in organizations that take user assistance seriously (as opposed to just chopping up a textbook and calling it help, which too many organizations still do) user assistance tends to be treated separately from the core documentation, which is still written in traditional textbook form. Switching to Every Page is Page One topics means moving to a user-assistance model for the bulk of the documentation.

The move to Every Page is Page One, and thus to a user-assistance style of writing, does not mean abandoning the attempt to educate the user. If anything, it means moving to a style that has proven to be more effective in educating people. As John Carroll's experiments showed, users don't learn well from books designed as systematic instruction. They learn by doing, making mistakes, and correcting them. They don't read systematically, but follow their own agenda driven by their current needs or interests. Interestingly, Carroll found that even those people who believed that they were systematic learners turned out not to be.

---

[1] Some user assistance is embedded in the interface. Such topics are different from regular EPPO topics, in part because the context is already established and in part because of constraints on length. While valuable, embedded help is tied to an individual feature, and the writer may not have the scope to consider different task contexts. Therefore I don't see it as a full replacement for conventional user assistance except in simple devices where the task context is clear. The best approach is to embed links from embedded topics to EPPO topics along lines of subject affinity.

> Most announced the conservative intention to read carefully before trying to do anything. Two even brought paper and pencil with which to take notes on their reading. None of the learners maintained this strategy for long. One apologetically informed us that we would have to watch him read both the *Owner's Guide* and the LisaProject book before he would use the system. Two minutes later, he had turned on the Profile (the hard disk drive unit). Another began the experiment by picking up the *Owner's Guide* and telling us that his style was first to read everything thoroughly. He actually read the manual for less than 9 minutes before switching his attention to the system and next referred to the manual almost 2 hours later.
>
> —*The Nurnberg Funnel*[8, p. 52]

Not only is it hard for writers to give up the textbook model, even users believe that they like and use it when actually they don't. Carroll even found that some of his subjects felt unsettled by the lack of hierarchical structure in some materials, even when they did better on the non-hierarchical materials.

This, of course, was in the 1980s, before most people had seen any kind of interactive information system, let alone the Web. The lesson that people can be unsettled by an unfamiliar organization of content is not to be neglected, but it has to be applied today in the cultural context of the Web where navigating by search and linking is now commonplace – for many, more commonplace than using a hierarchical TOC. To be sure, our cultural conditioning to hierarchy is still a factor, but it is breaking down as people become more experienced in non-hierarchical media.

Adopting a user-assistance approach to content, then, is not to walk away from educating users, but to walk towards an approach to education that encourages and supports what people do anyway (even if they don't think they do), which is to set their own curriculum. It is not even as if this were a preference or a choice. Rather, it stems from the fact that their current picture of the world is more real to them than anything they are reading, and it takes real world experience to shift that picture. As Carroll wrote (emphasis his):

> The problem is not that people cannot follow simple steps; it is that they do not. People are thrown into action (Winograd and Flores, 1986) they can understand only through the effectiveness of their actions in the world. People are situated in a world more real to them than a series of steps (Suchman, 1987), a world that provides rich context and convention for everything they do. People are always already trying things out, thinking things through, trying to relate what they already know to what is going on, recovering from errors. In a word, *they are too busy learning to make much use of the instruction*. This is the paradox of sense-making (Carroll and Rosson, 1987).
>
> —*The Nurnberg Funnel*[8, p. 74]

We teach best when we don't impose a curriculum, or even suggest one, but instead support each reader's unique path through our content.

Writing topics

Readers decide when and in what order they will read your topics. And their reading will be interspersed with other activities, including reading other people's content. And readers will read your topics individually, not as parts of a documentation set. Therefore, the right way to write topics is one at a time. Of course, documentation managers must take a broader view, but as a writer, you should try to write one topic at a time, focusing on the topic you are writing and its specific and limited purpose. The best way to do this is to focus on the characteristics of Every Page is Page One topics, which we covered in Part II. Let's look at how each of these characteristics can guide your writing.

Topics are self-contained

An EPPO topic is self-contained, which means writing an EPPO topic is also self-contained, in the sense that when you are writing, you should be focused on that topic and that topic alone.

This means that you should avoid switching back and forth from one topic to another as you write and that you should have a self-contained plan and objective for the topic you are writing.

There are a couple of reasons to avoid switching back and forth between topics. The first is that switching tasks creates cognitive overhead and makes it difficult to get into, and stay in, a state of flow, which is necessary to effectively and efficiently complete an intellectually demanding task.[2]

The second reason is to avoid unconsciously thinking of the topic as parts of a larger whole. Certainly you need to plan your topic set, but when you are actually writing an Every Page is Page One topic, it is best to approach it as an independent topic designed to fulfill a specific purpose for a qualified reader. This topic may be all that a qualified reader needs to accomplish that purpose, so you should write it with a similar degree of independence.

Create a plan for each topic.

To achieve this degree of independence, it is useful to create a separate plan for each topic. Obviously, you don't need to do an extensive planning exercise for every topic. You need a plan that is commensurate in scale to size of the task. But at the very least, your plan should state the specific and limited purpose of the topic and the topic type. Another excellent planning aid is to write the metadata for the topic before you write the topic itself (I will discuss this further in Chapter 19).

Topics have a specific and limited purpose

If you can only keep one characteristic of EPPO design in mind as you are writing, it should be this: define the specific and limited purpose your topic is meant to serve. Remember that your topic is an aid to the performance of a task and that a task is not simply a procedure.

Limiting the purpose of your topic is just as important as specifying the purpose. It is valuable to document the obvious extensions of the current subject that this topic is not going to cover. This helps to keep you from wandering off into those side issues as you write. Often when we write, our minds become engaged with the implications and side issues that arise from what we have just written, especially if, as is often the case, the act of writing has caused us to realize something new about the subject. Our pen can easily follow our mind down this newly discovered rabbit hole. We need clear boundary markers to remind us to stick to our defined purpose.

[2] See http://en.wikipedia.org/wiki/Flow_(psychology) for more on flow.

Of course, those ideas should not be lost. They may indicate a subject affinity that you will need to address in the broader documentation set. They should be recorded and fed back into the topic creation pipeline. Jotting them down has the dual advantage that it feeds the planning process and satisfies your mind's desire to capture the idea, freeing it to focus back on the task at hand.

The other virtue of explicitly stating the limits of the topic's purpose is that it helps you make sure you really do have a definite purpose in mind. Sometimes what seems like a clear statement of purpose can turn out to be amorphous and undefined when you set out to execute it. Documenting the boundaries helps to ensure that there is something real and concrete to define a boundary around.

Again, keep it simple and brief. You don't want the planning to overwhelm the writing, just provide sufficient guidance to keep the writing from running off the rails.

Topics conform to a type

Working with a good set of type definitions can really help in planning and executing a topic. A well-defined type will eliminate most of the planning required for topics of that type. The topic type is itself a plan for a topic of that type.

For instance, if you were writing a configuration topic for the embedded system mentioned in Chapter 9, you already know you need sections on Understanding, Planning, Configuring, Building, and Packaging. You also know that the Planning section consists of a series of questions for the user to answer, and guidance on how to answer them, and that the Configuring section consists of a formal flow diagram and a list of inputs and outputs. Knowing these things means a lot of your planning work is done and you can go right into the research phase.

Once you have defined the specific and limited purpose of a topic, you have defined the basis for a topic type. To fulfill a particular type of purpose – preparing a dish, using an API, configuring a module of an embedded system – you will need the same pieces of information for each instance of that type. The set of information fields required for each purpose constitutes the type of a topic on that subject. In order to make sure you cover your topic thoroughly and correctly – that you fulfill its specific purpose – you need to make sure that you include all the required information. This means you need to make sure your topic is true to its type.

Topic types can be defined and expressed in numerous ways. Wikipedia is full of topic types, but these do not appear to be the result of top-down dictation. Rather, the topic types develop over time as people fill in gaps in content and edit articles for conformance to the evolving standard for that class of content. You may not have time for that kind of evolutionary process, so you may need to take a more structured and deliberate approach to defining topic types.

This approach can take many forms. The topic type could be written into a style guide; it could be captured as a template file in FrameMaker, Word, or a wiki; or it could be defined in a database table or XML schema. The more formal the method and the more immediate guidance it provides to authors, the better compliance you will achieve. In my experience, people greatly overestimate their ability to comply with a type that is written in a style guide or template. Once a formal data model is created and the existing content is migrated to that model, people are amazed at how much was missing from their original content.

The more formal the method and immediate the guidance, the better compliance you will achieve.

The main thing in defining your topic types, however, is simply to focus on what information is required to meet the specific and limited purpose of the topic type. Avoid dragging general style issues or discussions about table formats or the like into the discussion of topic types. They will derail your topic type discussion every time. Settle those issues elsewhere or simply adopt an existing standard that covers them.

Often, the easiest way to design a topic type is to study multiple topics from different sources and look for what they have in common. If one source contains information not found in other sources, apply the specific and limited purpose test to it. If it is outside the specific and limited purpose of the topic type, reject it. If it is within the specific and limited purpose, include it.

Remember that this exercise is not about coming up with the one topic type that will encompass all the examples you are looking it. Many of your examples will include too much or too little. Instead, use the examples to help you figure out what is really essential to the specific and limited purpose of the topic. You may end up having to edit every one of your examples to fit the model you create, but they will all be the better for it.

Broadening your topic types to accommodate related material can make them harder to follow and audit.

Also be careful not to start broadening your topic types to accommodate related material. This can lead to a less-structured topic type that will be harder to follow and harder to audit against. If you think an existing topic model might work for some new content, go through the topic modeling exercise separately for the new content and then compare the models you produce. If they are equivalent, you can unite them. If not the differences are probably are there for a reason and reflect a difference in the specific and limited purposes of the two topic types. If so, keep the types separate.

When writing a new topic, determine which of your established types fits its defined purpose and follow it. If you find yourself wanting to include material that does not fit the model, ask yourself the following questions:

- Have I correctly defined and limited the purpose of this topic? (Chapter 8)
- Have I correctly identified the topic type that supports that purpose? (Chapter 9)
- Am I attempting to change levels within my topic because I fear the reader might not understand something? (Chapter 12)
- Am I trying to fit the subject matter of two topics into one or divide the subject matter of one topic into two?
- Have I discovered a special, weird edge case? These do happen, and the best solution is often to write a generic, untyped topic to cover them rather than adding support for every edge case to the normal topic type. This keeps the normal topic type simple to understand, follow, and audit.

If you exhaust all those possibilities, record it as a bug in the topic type definition. If your topic types are defined in a structured writing system, use a generic topic type for your topic until the topic type is fixed. Always note that you have deviated from the model and why.

If you define the specific and limited purpose for your topic and then find that there is no topic type defined for it, record the fact that a new topic type is required and write your topic as a generic type, trying as much as you can to develop a topic type as you go. The material you create will be an important ingredient in the topic type definition process for this new topic type.

Topics establish their context

Your working assumption for every topic should be that readers will arrive at the topic by searching, following a link, looking something up in the TOC or index, or following some other direct path. They will go straight to this topic without reading anything else. This is their page one. Therefore the first thing to do is to help them figure out if they are actually in the right place. To do this, the topic needs to establish its context.

If your information set has any form of local navigation, it is easy to assume that all readers will use that navigation to get to the content, and that by following that navigation they will have figured out the context of the material before they arrive there. This is not a safe assumption. First, there is no guarantee that people will use your navigation. People are increasingly search-dominant in their information seeking behavior,[3] so chances are they will arrive via search. Because people are not always that good at searching and because search engines are not always good at expressing the context of the results they return, people often arrive at the wrong topic and end up confused. Your topic, therefore, must orient the reader, just as page one of any document must do.

The key thing to remember about context setting, though, is that it's all about the subject matter. This is not "Welcome to the Acme Widget documentation set," nor is it "Congratulations on your purchase of an Acme Widget." It is not about context in the documentation set. It is about letting readers know which part of the real world this topic documents. Context setting defines what this topic is about and where its subject fits in the world.

The context of a topic in the real world and in the user's task space is also made explicit by the topic's metadata (more on metadata in Chapter 19). Showing topic metadata, particularly in a sidebar or as frame around the text, is a useful technique. See the examination of the Wikipedia article on the Manicouagan crater and *All About Birds*

[3] http://www.nngroup.com/articles/incompetent-search-skills/

in Chapter 4 for a discussion of this approach. However, if you are planning to reuse your content in multiple locations or multiple media, don't rely on metadata alone to establish context. Some media cannot display metadata in a convenient way or at all. Use the first paragraph of the text itself to set the context.

Make your context-setting paragraph brief and succinct. Don't fall into the trap of thinking you have to explain all the subjects you mention in the context-setting paragraph. Keep in mind the limits you have defined for this topic. However, go ahead and link from those subjects to topics that cover them.

Topics assume the reader is qualified

Like staying on one level, to which it is closely related, assuming the reader is qualified can be a difficult adaptation for an author to make. A well-defined and limited purpose can help, and a well-defined topic type can help even more.

For example, Figure 10.2, "Context-setting example" (p. 118), includes the sentence "App Engine's infrastructure takes care of all of the distribution, replication and load balancing of data behind a simple API – and you get a powerful query engine…" This statement assumes readers know what distribution, replication, load balancing, an API, and a query engine are. The qualified reader here is someone working in data-driven Web application design, and someone working in that field should be expected to know these things.

Of course, there may be readers who do not know what some of these things are. If they have only worked on small systems, they may not be familiar with load-balancing or replication. But because the topic established its context, readers are alerted at once that they might need to learn something new before tackling this subject. All that is missing are links to appropriate resources. Since these are general concepts, not specific to Google, readers could find resources using search, but that would be more work for them and wouldn't encourage them to stay in Google's content set. A better strategy would be to supply links to sources inside the content set.

Remember also that you are defining an assumption, not a real individual or a lowest-common-denominator. Many readers will come to your topic unqualified in some way or another. The world is not made up of cadres of identically prepared workers. Each unqualified reader is unique, and you can't write a topic that anticipates every possible way someone might be unqualified. Instead, write for the qualified reader – for the person who does this type of task every day – and provide links along the lines of subject affinity to help unqualified readers fill in the gaps.

You can't write a topic that anticipates every possible way someone might be unqualified.

This is why you need to record the subject affinities you find in your content. Those affinities help you manage the topic-creation pipeline to make sure your unqualified readers are provided for.

A big part of defining topic types is determining who the qualified reader is. Your topic type definition should tell you what qualifications to assume, and you should think about who the qualified reader is before you begin to write each topic.

Remember also that the definition of a qualified reader differs from one topic to the next. In the book-based documentation process, the qualifications of the reader were often defined at a single level for the entire documentation set, which is wholly unrealistic. In other cases, different levels of qualification were assumed, but only in terms of broad classifications such as novice, intermediate, and expert.

People's real qualifications don't fit on such a simple scale. The set of qualifications you need to understand and perform a particular task are quite specific and different from those of the next task. Therefore, when you picture the qualified reader for your topic, you need to define that reader in terms of specific knowledge and experience related to the subject of the topic.

Topics stay on one level

If you are used to writing books, you are used to changing levels. If you are good at writing books, you carefully plan how and when to change levels. As you plan changes of level you are trying to determine the best curriculum design to develop the reader's understanding of the topic.

Every Page is Page One, on the other hand, concedes that the process of acquiring understanding is not one we can successfully model or plan, and certainly not one we can generalize to an entire population of readers. Readers will change levels when they are good and ready, and thus EPPO topics do not attempt to impose level changes. They stay on one level.

The process of acquiring understanding is not one we can successfully model or plan and not one we can generalize to an entire population of readers.

The fact that topics differ so radically from books in this respect is one of the hardest things for writers to adapt to when they switch from writing books to writing topics. When they touch on a detail, their instinct as a book writer is to delve down into it on the spot.

When you switch to writing EPPO topics, you have to retrain that instinct. This does not mean suppressing it entirely. Those moments where you recognize the opportunity for a change of level or the chance to help readers understand a background concept are just as important when writing EPPO topics as they are when writing books. We just handle them differently.

In a book, the author decides whether or not to change levels; in EPPO, the author provides the means for readers to decide for themselves. Such moments always occur at points of subject affinity.

A good example is the instruction to "sweat the onions" in the recipe for Tarragon Mac and Cheese. The method for sweating onions is outside the specific and limited purpose of this recipe. However, the author might recognize at this point that not every reader will know how to sweat onions. The author of a book would then have to decide whether to change levels by including the method for sweating onions or to assume the reader can find out elsewhere and move on.

For the writer of an EPPO topic, however, that same recognition is handled in a different way. There is no question of changing levels – that would violate the specific and limited purpose of the topic – but there is an opportunity to observe and record the subject affinity and perhaps to act on it by providing a link.

Depending on your authoring process, writers may need to find and link to a resource themselves, or they may use semantic markup to record the subject affinity and leave it to the software to find a link – a process described in Chapter 20. In either case, it is useful to record the subject affinity in the topic source or separately if your authoring environment provides some method to do this. The list of subject affinities collected while developing content is invaluable to planning and managing your topic set. No matter how thoroughly you may think you have planned your topic set, the systematic management and recording of subject affinities will reveal many topics you have missed in your plan, often very important ones.

Topics link richly

Precisely because they are self-contained and conform to a type that expresses a specific and limited purpose, good EPPO topics need to link richly to other content to give readers easy access to any ancillary information they might need.

Linking in an EPPO topic should not be done selectively or on a hunch. It is fundamental to the bottom-up organization of topics and, therefore, something that should be done systematically. As noted above, managing and recording the subject affinities that you touch on in a topic is an important part of managing the writing process. EPPO topics link along lines of subject affinity, so the process of providing rich linking is really based on the practice of noting and managing subject affinities within a topic.

Chapter 20 describes a method for managing subject affinities and generating linking from them, but if that method is not available in your tool set, you still need to develop a method to record and manage subject affinities and generate links based on them.

From the writer's point of view, then, the essential principle is not to throw in a lot of links without regard to their value, but to systematically note and manage the subject affinities in your topics.

The question of style

It has become almost an axiom of topic-based writing that all writers who are writing topics must conform to a single style. There are both macro and micro reasons for this. The micro reason is that many topic-based writing systems, particularly in technical communication, focus on using topics as building blocks to create multiple publications. Naturally, if you want each publication created from building blocks to sound like it came from a single pen, every writer needs to sound the same.

Each topic is an information product in its own right, and readers traversing the web don't expect a uniform style across topics.

This concern is orthogonal to writing Every Page is Page One topics. There is no microscale reason why the style of one EPPO topic in a domain needs to be identical to the style of other EPPO topics in that domain. Every page is page one, so there is no expectation of continuity between one topic and the next. In many cases, the reader will be reading other topics in other domains between visits to your topics. Indeed, as readers travel around the Web and encounter frequent changes of style, they become inured to all but the most egregiously bad style. Each topic is an information product in its own right, and readers traversing the web don't expect a uniform style.

Therefore, unless you are planning to build EPPO topics out of smaller building block topics, there is no micro-level reason to worry about uniformity of style.

The macro-level reason for uniformity of style is a marketing-driven desire to have all of a corporation's content conform to a certain voice and tone. If this is the corporate standard, you will naturally need to comply. But it is worth noting some reasons why a corporation might be well advised to take a different stance.

Consider these findings from "What every blog needs to be great"[4] by Jesse Stanchak:

> SmartPulse – our weekly reader poll in SmartBrief on Social Media – tracks feedback from leading marketers about social-media practices and issues.
>
> Last week's poll question: **Which of these qualities is most important to a blog's success?**
>
> - A distinctive voice – 43.41%
> - Compelling exclusive content – 35.66%
> - A unique niche – 11.63%
> - Strong promotion via social-media channels – 5.04%
> - Excellent SEO – 2.71%
> - Connections to famous brands, personalities – 1.55%
>
> —Jesse Stanchak, editor of SmartBrief on Social Media, writing at SmartBlogs.com

[4] http://smartblogs.com/social-media/2010/09/01/what-every-blog-needs-to-be-great/

43.41% of respondents chose a distinctive voice, even over compelling and exclusive content. Now this is one small survey, specific to blogs, but there are broader reasons to believe in the power of a distinctive voice on the Web today. David Weinberger's observations that the Web gives us access to experience as well as authority and that we now give our trust to our social networks more than to institutions suggest that the bland and anonymous corporate tone may not be the most appealing choice in every case. Even notoriously buttoned down and conformist Apple essentially built its brand on the personality of one man: Steve Jobs.

Topics with a distinct style and tone often stand out and, therefore, are more likely to be filtered into the set of topics a reader chooses. This is even more so when the topic comes from an identified person that the reader knows and trusts. If you are trying to build social features into your documentation set, it is unreasonable to expect a community to develop around a bland and anonymous documentation set. People are attracted to people. Put the author's name and photo on each EPPO topic, and you will have a much better foundation for building a community around your documentation. Atlassian is an example of a company that puts the name (though not photo) of the author on each documentation topic.

Concerning reference information

In Chapter 8, I compared a documentation set to a bus system. Topics provide transportation between known hubs. Of course, your network won't be complete or perfect. It won't cater to every journey. Some routes aren't traveled often enough to make regular service economical. Sometimes readers will have to find their own way.

Readers are not entirely stuck if you don't give them a topic specific to their purpose. People are, to varying degrees, smart, brave, determined, and patient, and if they can scrape together enough facts, they can construct a method for themselves. In fact, some among your user community will positively thrive on the challenge (and, as a bonus, they will often leave a map of the route they have taken for others to follow, contributing to the long tail).

But even the bold and the brave are at a loss if they can't get at the facts. Reference information is the foundation on which an effective documentation set is built.

Reference information is the foundation on which an effective documentation set is built.

This, then, is what the ideal documentation set should look like: a foundation of a solid, well-organized database of facts upon which is built a highly navigable network of sensibly sized topics connecting sensible hubs.

In an EPPO information set, references play a special role. Because EPPO information sets are link-based, topics frequently have occasion to link directly into references. A reference, then, is a rich link target in an EPPO information set. The existence of a solid foundation of reference content makes it easier for topics to stay on one level and stick to their specific and limited purpose.

Concerning tutorials

A tutorial is systematic instruction, and, as John Carroll demonstrated, systematic instruction often does not work well. On the other hand (and as Carroll also discovered) readers are not always realistic about what they want, and they often ask for tutorials. Are tutorials compatible with the Every Page is Page model?

If you are expecting the reader to sit down and complete several hours of tutorial, then the answer is probably no. You have then moved into the realm of self-directed training, which is a discipline in its own right. But if you have the more realistic expectation that readers will probably only follow the tutorial for a few minutes before striking off on their own, then the answer is yes.

The immediate implication of creating Every Page is Page One tutorials, though, is that even if you have several of them, and even if they are of increasing complexity, they should not be written as if users will study the whole series in order. Like any EPPO topic, EPPO tutorials should be written with the assumption that readers will be qualified to take whichever tutorial they choose. EPPO tutorials should establish their context so readers can determine if they are qualified and provide links to material readers can use if they need to become qualified.

As a practical matter, tutorials are often written in a sequence where the output you create in performing one tutorial is the input you need to do the next tutorial. That can be dealt with by providing prefabricated inputs separately with each tutorial.

Like any other topic type, a tutorial is full of subject affinities, and providing links along those lines of subject affinities can help users fill in any gaps in their knowledge as they work through the tutorial. A tutorial topic, in other words, should be written like any EPPO topic. It should be self-contained, have a specific and limited purpose, conform to a type, stay on one level, assume the reader is qualified, and link richly.

Concerning videos

Video is becoming an increasingly important part of technical communications. It is often users who are taking the lead and producing their own how-to videos on YouTube, but professional technical communicators are catching on, too. How-to videos are almost always Every Page is Page One in nature.

Unless they are part of a planned video curriculum, such as on lynda.com, videos are usually self-contained. It is perhaps even more important with a video than with text to make sure that the video is self-contained and functions independently. Because videos can't contain links and don't support skipping and skimming, your users can't construct their own curriculum, at least for the duration of the video. Therefore, a video needs to work as a whole.

A good video always starts by establishing its context. In this respect, videos sometimes have an advantage over text because they can use multiple media to establish context. As always with video, you should think first in terms of visual presentation when you establish context rather than narration.

Videos generally assume the viewer is qualified to watch, since backtracking and asides are even more distracting and annoying in a video than in text.

One of the interesting properties of videos is that they are significantly harder to edit than text. Video editing is almost always something you do to assemble multiple initial takes into a finished video. It is part of the original production process of the video, just as copy editing is part of the original production process of text. Once a video is complete, it is very unlikely that someone will come along later and make significant revisions or add new material. The location, the lighting, the narrator's voice, the production levels, even the flow of the narrative are so hard to reproduce that it is usually easier to shoot a new video than to revise an old one.

One side benefit of this is that it helps to keep videos on one level. In many cases, a text that hops and skips from one level to another is the result of revision over time by many different writers (often with different levels of knowledge and skill and differing amounts of time available for the task).

Another factor that works to keep videos on one level is that different levels of abstraction tend to require different visual styles. How-to information usually involves pictures of the physical device or screen you are discussing, while more abstract material is generally better presented using analogies which may take the form of animations or pictures of real-world objects.

Full-length documentaries, of course, can use several visual styles and visual metaphors and can change levels to tell a complete story. Like the author of a book, the producer of a full-length documentary decides when to change levels. But in an EPPO video, as in a written EPPO topic, it is best to stay on one level and let the reader choose whether and when to change levels by selecting a different topic or video.

Videos and linking

One of the chief problem of video, as compared to text, is that there is no good way to embed inline linking in a video. It is not technically impossible to insert a link into a video, but it does not work very well. A link provides an opportunity for the reader to pause and then decide whether to continue on the current topic, follow some subject affinity, or move to a higher or lower level. That consideration requires a momentary pause on the viewer's part, and video players are not yet intuitive enough to know when the viewer is contemplating a move to another topic. (At time of writing, there is talk in the rumor mill of phones that will use their camera to watch your eyes and pause a movie if you look away, so pausing when you consider following a link may not be such a far-fetched idea by the time you read this.)

So, for the time being at least, most videos don't contain links. But videos are often embedded in pages that do link richly, as anyone who has used YouTube knows. Indeed, YouTube would not be what it is today if it were just a video player. What makes

YouTube work are the targeted links to other videos that are provided as a frame around the current video. These links include similar videos, videos by the same uploader, videos by the same artist, and videos similar to videos you have viewed in the past. Indeed, watching YouTube videos is rather like eating popcorn, you can't stop at just one. Rather, you are likely to end up curating and watching your own personal film festival.

Therefore, when you use videos in your documentation, consider putting them in a frame so you can include links to related subjects.

Videos as topics

Since videos share so much in common with Every Page is Page One topics, it makes sense to treat them as topics for purposes of organization. While you can certainly provide a list of videos if you want, it does not make sense to segregate videos into their own little ghetto, separate from text topics. Videos are good for conveying certain kinds of content, and text topics are good for conveying other kinds of content. As users navigate through the content set, it is entirely appropriate for them to encounter each piece of content in the most appropriate media for that content.

Because the means to view videos, particularly videos that rely on sound, may not always be available, you may wish to provide an alternative text topic covering the same material. This can be in the form of a video transcript, but if a video transcript can convey the entire meaning of the video, either you are not taking full advantage of the visual and auditory capabilities of the video medium, or your topic is simply not a good choice for video. So it makes better sense to provide a textual topic twin that is written as an EPPO topic.

The text and video twin topics should be kept together as much as possible. The simplest way to do this might be to embed the video topic inside the text topics. An alternative, if your publishing platform supports it, is to have the two link to each other, perhaps using a tabbed interface, so the reader/viewer can flip back and forth between the video and text versions of the topic.

In either case, you want the twin topics to show up as one in search results, and certainly you want to be sure that a reader/viewer who finds one in a search result has immediate access to the other. Twinning text and video topics like this can go a long way to solving the SEO and linking problems associated with video. It can also be an answer for viewers/readers who get impatient because they can't skip and skim a video. They can switch to the textual twin if they get too impatient to sit through the video to the end. And twin topics give you the opportunity to provide alternatives for readers who need more accessible content.

Videos as objects

Not every video is necessarily a topic on its own. Sometimes videos and animations are a great way to show a physical process or to explain a concept, just as a picture or a drawing may be. But just as pictures and drawings are not topics in their own right, such videos are not topics either. If a video is not Every Page is Page One by itself, then it needs to be embedded inside a regular EPPO topic (or a reference entry), just like a picture or a drawing.

CHAPTER 15

Every Page is Page One Topics and the Big Picture

Writers often ask how topic-based writing can handle the big picture. Topics, they complain, don't provide a way to tie everything together. Thus users get lost in a sea of topics, can't understand the system as a whole, and can't figure out where to start.

Every Page is Page One topics are a natural form when you write about an isolated subject or write in isolation about one aspect of a large subject. However, when you are charged with documenting a whole large subject, whether alone or with others, it can be difficult to figure out how to cover it all with Every Page is Page One topics. It may feel more natural to write a book rather than a set of discrete topics.

But just because the subject is large does not mean readers will want to read about all of it at once. Readers usually want to read only about the specific aspects of the subject that affect their current task.

This is not to say that readers have no interest in learning. It is simply that most are not interested in sitting down and learning everything before they do anything. Indeed, most would prefer to learn as much as possible by doing, turning to documentation only when they are stuck and have no other recourse. (It could be argued that we learn everything by doing and that content only informs the doing.)

Even in regulated and safety-critical industries, where learning by doing cannot be tolerated for reasons of safety and security, employees are seldom expected to learn everything up front by reading. Simulators and one-on-one training are used to support learning by doing without the danger of breaking things. There are defined steps up the ladder of responsibility where information gained from content or instruction is reinforced by supervised practice.

Books and the big picture

Many product manuals do a lousy job of giving the user the big picture. Consider how most manuals handle the big picture. You will rarely find a chapter devoted to the big picture. Academic practice would have us lead with the theoretical overview. But then minimalism rebukes us and insists that we must get right to the action. What to do with the overview then? It would seem silly to put it at the end or squeeze it into the middle somewhere, so out it goes.

How then to deal with the big picture? What generally seems to happen is that it gets woven into the chapter structure of the book. As readers progress through their sequential reading of the book, they are expected to gradually pick up the big picture, while constantly being kept busy and entertained by all the practical hands-on stuff they are reading/doing along the way. The big picture is not expressed explicitly; it is to be inferred from the sequence of the book.

Many defenders of the traditional table of contents make exactly this claim: that the table of contents gives the reader the big picture. It is not only the curriculum, it is the grand summation of the subject matter. No wonder then that writers feel that they need to string topics together in a prescribed sequence in order to give users the big picture. That is how they are used to doing it in books.

Even in the book world, having the TOC or, more generally, the order of chapters carry part of the meaning of the work seems to me like poor information design. For it to work at all, readers have to actually read the TOC and the book in order and in a fairly short period of time. Even if they do all of that, I'm not convinced that the typical reader would recognize or grasp the meaning implied by the order of the work. A table of contents is an instrument of navigation, not a carrier of meaning.

In a help system or on the Web, I'm certain that any information or meaning implied by the order of topics, whether linear or hierarchical, is going to be lost on the reader, who is often accessing the content at random.

Technical communications, in any case, is not a field where meaning should be left to implication. If there is some information that the reader needs, it should be explicitly stated, not implied by a TOC or by the order of chapters.

The priority of the big picture

The Every Page is Page One approach is not in any way a rejection of the need to provide the big picture. Rather, it views the big picture as an important task enabler that users will seek when they attempt a task that requires knowledge of the big picture. But the Every Page is Page One approach also acknowledges that until readers want the big picture, they won't seek it out, and there is no point trying to force it on them before they are ready.

Some may object that users often waste a lot of time and effort trying to do things the wrong way because they do not grasp the big picture. This certainly happens, but it is equally true that you cannot instruct users until such time as they deign to look at the documentation. Constructing the documentation as a textbook that assumes users will study the whole book before acting – and then blaming them for their failures when they neglect to do so – is not an effective strategy. We know this is not how people behave. We need to provide information in a form that works with the way users actually behave.

On this score, we can harken back to the idea that the main task of documentation is decision support, which we discussed in Chapter 8. We know that the odds of our users reading the textbook explanation of the big picture is slight. But if we document individual tasks with a decision support focus, the ways in which the big picture affects the performance of those tasks will always be present. And if the task topic is richly linked, as it should be, it will link to the big-picture topic.

We need to provide information in a form that works with the way users behave.

Few people start with the big picture. Without practical experience, the big picture is an abstraction that it is difficult to fit into one's view of the problem space. The desire for the big picture generally arises from the desire to make sense of specific concrete experiences. When users get stuck or stumble attempting a specific low-level operation, that is when they turn to the documentation. Then, by showing its relevance to the task, the writer can suggest the value of understanding the big picture.

Writing the big-picture topic

The job of the big-picture topic is to give the big picture without delving too deep into the details. It is not an overview of a book or a curriculum, it is the 10,000-foot view of the subject. Like any EPPO topic, it should be self-contained and stick to its level.

For an example of a good big-picture topic, check out "What Is Google App Engine?"[1] in the Google App Engine docs. Tellingly, on the App Engine home page this topic is not presented as a starting point. It is listed under the heading Dive Deeper, as shown in Figure 15.1. (Curiously, the title of the link, "App Engine Basics," does not match the title of the article itself, "What is Google App Engine.")

Whoever designed this article clearly knew minimalism because the Get Started column is all about getting some initial experience. It assumes that the desire for a big-picture topic will come later, when the reader is ready for a deep dive into the subject, and that is where it is placed.

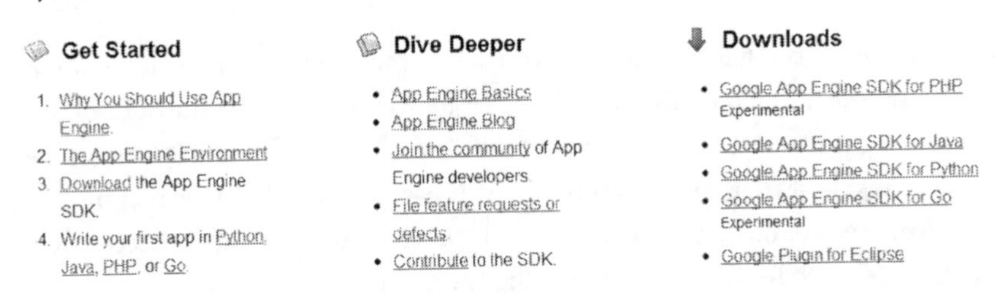

Figure 15.1 – Location of App Engine Basics topic on the App Engine home page

Big picture topics tend to be long, and this one is no exception, so I can't reproduce it here, but an outline and a couple of highlights should suffice to show how it works.

[1] https://developers.google.com/appengine/docs/whatisgoogleappengine

What Is Google App Engine?
- The Application Environment
  - The Sandbox
  - The PHP Runtime Environment
  - The Java Runtime Environment
  - The Python Runtime Environment
  - The Go Runtime Environment
  - Storing Your Data
    - The Datastore
  - Google Accounts
  - App Engine Services
    - URL Fetch
    - Mail
    - Memcache
    - Image Manipulation
  - Scheduled Tasks and Task Queues
- Development Workflow
- Quotas and Limits
- For More Information ...

Figure 15.2 – Outline of What Is Google App Engine?

Figure 15.2 shows the outline of the topic. In some ways, it looks like the table of contents of a book, but the topic is nowhere near the length of a book and stays firmly at the top level. It never dives down to lower levels, but it links to them as needed.

> The Python environment provides rich Python APIs for the datastore, Google Accounts, URL fetch, and email services. App Engine also provides a simple Python web application framework called webapp2 to make it easy to start building applications.

Figure 15.3 – Detail from What Is Google App Engine?

The section shown in Figure 15.3 talks about the Python APIs for accessing the datastore, Google Accounts, etc, but it does not delve into the APIs themselves. The big picture is that the APIs exist and that they address major areas of concern. The detail of what is in the APIs and how to use them is not part of the big picture. It is easy to imagine a traditional user manual for the App Engine diving down immediately into these APIs. An EPPO big-picture topic won't change level and, therefore, will make it easier for readers to grasp the big picture whenever they are ready for it.

Finding the end of the string

Getting the big picture of a complex product is not something that happens in a few minutes of reading. We don't learn that way. We build up a big picture over time, through experience and exposure. No matter how good your big-picture topic is, no reader is going to read it through and immediately understand the big picture. As much as anything, the role of the big-picture topic in a properly linked topic set is navigational. It helps users find the area of the product they need to focus on and provides rapid access to the topics that describe that area.

Readers who are looking for the big picture are not always looking for the grand overview. They are really after the end of the string. They want a starting point.

What most users want is a way to get going. They don't want the big picture. They just want to find the door marked Enter.

This, of course, is the need that the venerable *Getting Started Guide* is designed to fill. But Getting Started guides tend to be tutorials built on the assumption that users want to be taught and that all users are looking for the end of the same piece of string. Neither assumption is warranted in light of what we know about user behavior.

What most users want is a way to get going. They don't want the whole big picture. They just want to find the door marked Enter. But each user is looking for a different door; each user has a different background, a different set of assumptions, a different picture of how the world works, and a different purpose. Because of this, the big-picture topic can play a second role, that of being a room full of doors. In this case, even if a user doesn't grasp the entire big picture, the big-picture topic still provides the context needed to select the right door. Therefore, a big-picture topic may be a place that readers arrive at from the bottom up or a place where they start exploring content that interests them.

I'm not suggesting that you should get rid of the Getting Started Guide. The words "Getting Started" are familiar and comforting to readers, and that alone is justification for retaining those words, though the word "Guide" can safely be dropped since we are writing topics, not guides. However, a Getting Started topic should do what minimalism always prescribed, which is to avoid artificial tutorials and get readers started on real-world work as soon as possible.

Pathfinder topics

One step below the big picture – and in line with the idea that different readers are looking to get started on their own projects, not on toy learning projects – there is frequently a need for what I call *pathfinder* topics.

A pathfinder topic shows the reader the overall path for accomplishing some real goal with your product. It is not a beginner topic because it does not assume that users want to do something very simple for practice. Instead, a pathfinder topic covers the full range of tasks and features in a way that helps the user get a grip on how to attack a problem. But it gives none of the details. The details are left to the task-oriented topics that make up the bulk of the information set. Good Every Page is Page One topics stay on one level. Pathfinder topics fit at a level below the big-picture topic and above workflow or task topics.

A pathfinder topic shows the reader the overall path for accomplishing some real goal with your product.

The role of the pathfinder topic is to set the user's feet on the right course. In most cases, that is why users want a view of the big picture in the first place – they want to work out a plan of attack for a certain kind of problem. A straight-up big-picture topic is certainly worth having, but the real work of guiding users down the right path belongs to the pathfinder topic.

For an example of a pathfinder topic, check out the WordPress Codex topic titled "Photoblogs and Galleries."[7] Example 15.1 shows the beginning of the topic. It concludes with a long list of related resources (omitted here).

---

[7] http://codex.wordpress.org/Photoblogs_and_Galleries

Example 15.1 – Photoblogs and Galleries pathfinder topic

Photoblogs and Galleries

A picture says more than a thousand words.

Images can be displayed in many ways in WordPress, from simple usages to complex galleries, to even mostly wordless PhotoBlogs. Let's look at some of the different options for PhotoBlogs and Gallery scripts.

PhotoBlogs

PhotoBlogs are different from normal blogs. Normal blogs put the emphasis on the words with only the occasional image featured. PhotoBlogs are all about the images and not the words.

The easiest way to get started with photoblogging on WordPress is to either install Johannes Jarolim's YAPB (yet another photo blog) plugin or the PhotoQ Photoblog plugin.

YAPB adds all of the functionality of a standard photo blog directly to WordPress with a minimum of configuration. It includes automatic image resizing, exif data, and other tools. Johannes' site has links to themes already incorporating YAPB, otherwise you can make your own.

PhotoQ takes a slightly different approach, it gives you a queue which you can fill with photos to be posted on your photoblog. PhotoQ is geared to batch processing of photos and it features batch uploads, automatic image resizing, exif support, watermarking and automatic posting via cronjobs among other features. PhotoQ works with most WordPress themes without adaptation and includes an auto-configuration option for some of the most popular photo centric themes.

This topic is not a big-picture topic for WordPress as a whole. Nor does it provide specific instructions on how to perform any particular task. Rather, it addresses a fairly high-level subject: how to create a Photoblog or Gallery. It does so by walking the reader though the difference between a photoblog and regular blogs (context setting) and then discusses the various options and resources available. Readers will have to read other topics to get specific directions on using these tools (though they may just install them and learn by trial and error). But this topic helps readers understand what capabilities are available and start choosing how to proceed. In other words, it sets readers on the right path.

Pathfinder topics set readers' feet on the right path to achieving their goals.

CHAPTER 16

Sequence of Tasks vs. Sequence of Topics

One of the objections I often hear from writers is that Every Page is Page One's insistence on removing sequential dependencies makes it difficult to create a defined order of topics, for example when a set of topics forms a workflow. The question I ask in return is, if there is a workflow here, shouldn't you have a topic describing that workflow explicitly? A workflow is too important to be expressed only by the table of contents.

Following a workflow is an important part of using many products. If the workflow is only expressed through the sequencing of topics and is not otherwise documented, the only thing that expresses that workflow is the table of contents.

A workflow is too important to be expressed only by the table of contents.

That is certainly not a good thing on the web, where people will frequently land on your content as a result of a search and, therefore, will not have seen or used the table of contents. I don't think it's a good thing in an online help system, either, where the reader may come to the topic via the search or index rather than the TOC. And I suspect is not a good thing even in a book, where, even if they begin by browsing the table of contents, readers will probably not recognize that certain sections of the TOC describe a workflow.

One of the most common ways of representing a task sequence in a book is by a sequence of sections or chapters that matches the sequence of tasks. This is not an entirely satisfactory approach because many task workflows are more complex than a simple sequence, with branches, loops, and exceptions being common. Also, many tasks are part of more than one workflow, which leads either to duplication or awkward cross-referencing.

The Every Page is Page One solution to the sequence-of-tasks problem is to write a single, separate workflow topic that captures the overall sequence and points to the constituent tasks in the appropriate order.

This approach is preferable to using a fixed sequence of topics. It allows you to create multiple workflows, each of which may be more clearly expressed through a separate topic. This avoids some of the convoluted conditional expressions required if multiple workflow variants are expressed in a single linear topic sequence. It also avoids the need to reuse the task topics in multiple locations.

Therefore, no matter the medium, a workflow should always be described in a topic of its own. Since workflows generally consist of multiple tasks or procedures, a workflow topic can refer to those tasks and procedures and link to the topics that describe them.

This is not to say that in safety-critical situations you might not want to use a fixed linear form of instruction. There are cases where that is appropriate. A pilot checklist is one example. The shutdown procedure for a nuclear power plant is another. In these situations, a fixed linear order is appropriate. However, precisely because of the safety implications, such procedures should always be presented in one topic and not implied by the TOC.

I also fear another pitfall. Whenever information is implied – by a TOC, the order of chapters, or any other mechanism – there is a danger that the writer won't fully understand the big picture. By requiring writers to fully document a workflow that was formerly only implied, you may expose holes in their understanding. (It is said with reason that you never really learn something until you try to teach it.)

And finally, if important information is implied by the order of the document or the structure of the TOC, will reviewers notice this information or discover flaws? (Personal experience tells me that the answer to that question is a definite No.)

Working backwards

In a comment on my blog, Rebecca Hopkins raised an important issue.

> You may come into the help looking for information about creating letters. Since creating letters is a subroutine for many modules, if you don't already know the routine, you'll be working backwards: from how to create a case letter to how to create a case; from how to create a campaign letter to how to create a campaign. The ToC lets you work backwards that way.
>
> —Rebecca Hopkins, Comment on EPPO blog[1]

When we talk about using the TOC to work backwards through a help system, we could be talking about the common case where search lands you on a topic that is not an Every Page is Page One topic, is not self-contained, and does not help you by itself. In that case you need to work backwards through the TOC simply to find the beginning of some piece of content that is sufficiently self-contained to be useful.

But that is not what Hopkins is describing here. She is describing the problem of landing on a topic that may be self-contained, in the sense that it fully serves a definable purpose, but which nevertheless describes a task that is part of some larger project or projects. This frequently occurs when readers begin a task without reading the documentation, get stuck, and, only then, consult the documentation. Often, the reason they are stuck is not related to the task they were working on when they realized they were stuck, but to something they did earlier. At that point, what they need to do is walk backwards, not just through the content, but through the work they have been performing.

Sometimes the reason they are stuck is not related to the task they were working on when they got stuck, but to something done earlier.

TOCs are a poor vehicle for working backwards in this way. At best they can only support moving backwards along a single axis. But different readers may need to move backwards along different axes, looking for different prerequisite concepts or tasks. The TOC, which can only show one backward path, is forced to play favorites and support only one of many possible user reasons for moving backwards through

---

[1] http://everypageispageone.com/2013/02/19/a-new-approach-to-organizing-help/#comment-117378

the subject matter. A better approach is to have the topic itself provide the means to work backwards, particularly in the context-setting section.

> **Introduction:**
> This page documents the API (Application Programming Interface) hooks available to WordPress plugin developers, and how to use them.
>
> This article assumes you have already read Writing a Plugin, which gives an overview (and many details) of how to develop a plugin. This article is specifically about the API of "Hooks", also known as "Filters" and "Actions", that WordPress uses to set your plugin in motion.
>
> These hooks may also be used in themes, as described here.
>
> —WordPress Codex Plugin API[2]

Figure 16.1 – Working backwards using context

Figure 16.1, "Working backwards using context," shows a simple example of this from the WordPress codex. This topic is an introduction to the WordPress API, and in the context setting paragraphs it places the API in context as a means to write plugins or themes. Using the API is part of a larger project to create either a plugin or a theme, so the reader who drops straight into this topic may need to walk backwards to plugins or themes. This topic provides links to allow them to do just that.

By making a topic's context navigable, you support this kind of walking backwards through the subject matter.

---

[2] http://codex.wordpress.org/Plugin_API

CHAPTER 17

EPPO and Minimalism

One of the foundational ideas of Every Page is Page One information design comes directly from John Carroll's studies that led to the creation of minimalism – his observation that what he called the systematic approach to instruction, in which everything is laid out for the reader in systematic fashion – simply does not work. People won't follow the system.

> Learners also often skip over crucial material if it does not address their current task-oriented concerns or skip around among several manuals, composing their own ersatz instructional procedures on the fly.
>
> —*The Nurnberg Funnel*[8, p. 8]

EPPO as a platform for minimalism

Every Page is Page One is founded on the idea that people simply will not read linearly or sequentially, a fact confirmed by any number of studies of reader behavior on the Web,[1] as well as studies like Carroll's that showed the same behavior with paper manuals.[2] Every Page is Page One is designed to accommodate and facilitate this nonlinear reader behavior, rather than to resist or punish it. It does this in several ways:

---

[1] See, for example, http://www.nngroup.com/articles/how-users-read-on-the-web/, http://www2.parc.com/istl/groups/uir/publications/items/UIR-2001-07-Chi-CHI2001-InfoScentModel.pdf, and http://www.uie.com/articles/three_click_rule/.

[2] Occasions for linear reading, and the willingness to do it, certainly exist. They exist where the reader needs to understand a new idea before acting, where multiple threads of argument must be assembled to establish an idea, and in areas that are more abstract and don't present an opportunity to learn by tinkering. But for technical communications, expecting users to read and read and read without touching the product is just unreasonable. Even if they must do a big read to understand the product, users will only accept this requirement as the result of experience. The help content they consult in the meantime can point them towards the need for a big read, but that content still needs to be EPPO content.

- By making each topic self-contained, without reliance on previous or next topics, EPPO allows readers to choose any topic at any time.
- By having a specific and limited purpose, EPPO topics avoid extraneous material and focus on the reason readers came to them.
- By establishing their context, EPPO topics make it easy for readers to determine where they have arrived.
- By conforming to a type, EPPO topics make it easy for readers to skip and skim or hone in on one particular piece of data.
- By assuming readers are qualified, EPPO topics don't waste time on introductions that readers are not interested in.
- By linking richly along lines of subject affinity, EPPO topics help readers jump around in the subject area and find the information they want next.

These things are very similar to Carroll's own hypothetical prescription:

> Escaping these problems, and providing for material to be sensibly read in any order, necessitates a different approach to organizing instruction. It requires a high degree of modularity, a structure of small, self-contained units. The internal organization of the units must be streamlined so that learners are not likely to skip around within them; the organization among units must be simple so the learners can more successfully skip over or skip units. One cannot eliminate the need for prerequisite knowledge in learning, but one can try to minimize the tangles and side effects that occur when prerequisite relationships are disregarded. Attempting an advanced task "out of order" should motivate and help guide the acquisition of (prerequisite) skills that may have previously seemed irrelevant.
>
> —*The Nurnberg Funnel*[8, p. 85]

In the course of his experiments, Carroll created and tested a set of what he called "guided exploration cards." His design was, in most respects, that of Every Page is Page One topics.

> The organization of the deck of cards was intended to facilitate getting started fast; it was clear that only the first few cards needed to be used at all in order to get going. However, the cards were also designed to be used unordered. Each card attempted to address its specific procedural questions without reference to material covered on the other cards. Thus, (ideally) each card could be initially read, and later referred to, independently if all the other cards. We hoped that this would encourage and support opportunistic problem solving on the part of the learners.
>
> —*The Nurnberg Funnel*[8, pp. 115–118]

The cards were self-contained and not order-dependent.

> The cards were modular with respect to their internal organization of information. Each consisted of a goal statement, hints, checkpoints, and remedies. These four procedural components were graphically blocked off on the cards to stress the functional decomposition to learners.

Each card conformed to a defined type.

> The goal statement was a brief description of the subject matter of the card....

Each card began by establishing its context in the subject space.

> The cards tried to conform to what we expected the learners to know already and to some practices they might already have. For example, we tried systematically to exploit typewriter knowledge.

The cards assumed the reader was qualified and were written with a definition of the qualified reader in mind.[3]

[3] Even though Carroll's studies are pre-internet, they are relevant because they show how people used information-foraging behavior even in paper manuals.

Is EPPO minimalist?

Of course, there are principles in minimalism that are not required characteristics of EPPO topics. The cards that Carroll created provided hints rather than ordered procedural steps. EPPO is silent on this. You can write an EPPO topic that is minimalist in this respect or systematic. EPPO is not, in itself, minimalism and is open to the possibility that minimalism might not always be appropriate. However, EPPO does draw strong inspiration from the minimalism experiments, and it does provide a good platform on which to create a minimalist documentation set.

Some might see the larger structure of an EPPO topic as being less consistent with minimalism than a bare-bones procedure or task topic type. I don't think that is the case. For one thing, minimalism is not particularly keen on procedures, finding that people rarely follow them correctly. Therefore, while giving people only a procedure is minimal, it is not minimalist because it does not invite exploration or learning.

While giving people only a procedure is minimal, it is not minimalist since it does not invite exploration or learning.

The context-setting content that an EPPO topic contains may sound dangerously like the sort of front matter that minimalism recommends omitting. It isn't. Indeed, providing context is necessary to minimalism's goal of getting to action. You can't act effectively if you don't know where you are.

A key finding of the minimalism studies is that readers strike off on their own paths through information rather than sticking to the path created by the author. A key principle of EPPO is to create content that facilitates, rather than frustrates, readers in choosing their own paths. This does not mean that readers will always read an entire EPPO topic from beginning to end. Indeed, EPPO design recognizes that readers will often leave in the middle of a topic to switch temporarily to a higher- or lower-level topic. Readers may realize they are not qualified and need additional information, or they may decide they are not in the right place and need to find another topic. In recommending conformance to a well-defined type for every topic, EPPO also recognizes and accommodates the reader's self-directed exploration of a topic.

The point is not to find a single path that readers will always follow sequentially, but to optimize their overall transit through the information set. And trying to break content into units so small that it's impossible for a reader to leave a piece before reading the whole thing is not the best way to achieve this.

Readers seek material based on its subject matter, and even if they don't want to read everything on the subject at once, there comes a point where subdividing the subject matter further just makes it harder for readers to find what they want. For instance, a reader who is making a shopping list may just want the ingredient list of a recipe, but separating the ingredient list into a separate topic would not make it easier to find or to read. Quite the contrary, separating the list would make it more difficult to find, not to mention making it harder for the reader to cook the dish.

The point is not to find a unit readers will follow in sequence, but to optimize their transit through your information.

The reader's ability to immediately select the parts of a topic that are of interest is enhanced when the different pieces of information that make up a topic are clearly laid out and consistently related with each other. Far from getting in the way, these characteristics of an EPPO topic help readers locate and focus on the information they are looking for.

There is clearly a tension here between the number of objects readers have to search through to find the subject they are interested in and the amount of time it takes to locate a particular piece of information within one of those objects. Micro-level searching – scanning the immediate vicinity for clues – is efficient, but only if the immediate vicinity is small and well structured. On the other hand, splitting closely-related material into separate pieces can make macro-level searching more difficult – you have to find all the pieces, and the smaller pieces may be hard to identify. EPPO topic design is very much about the human scale of information.

A good EPPO topic is optimized for both micro-scale searching and for narrative reading. I think this better supports the general aim of minimalism: to make the documentation experience less obtrusive for the reader than either a book on the one hand or a Frankenbook on the other.

EPPO topic design is very much about the human scale of information.

Minimal vs. comprehensive

There is a kind of paradox in the minimalist approach. It seeks to free readers to strike their own path through the documentation, to encourage experimentation, and support error recovery. Yet in being less comprehensive, it runs the risk of not providing material that readers on their own path may want or material necessary to support error recognition and recovery. In the paper world, the sheer size of the documentation set could be a barrier to the users' freedom to set their own paths.

In the paper world, the sheer size of the documentation set could be a barrier to your users' freedom to set their own paths.

Carroll writes:

> The key idea in the minimalist approach is to present the smallest possible obstacle to learners' efforts, to accommodate, even to exploit, the learning strategies that cause problems for learners using systematic instructional materials. The goal is to let the learner get more out of the training experience by providing less overt training structure. The approach does not solve the learning paradox; rather it compromises in the direction of accommodating the learner's desire for meaningful interaction at the expense of providing a less comprehensive curriculum.
>
> —*The Nurnberg Funnel*[8, pp. 77–78]

In the book world, providing a less overt training structure comes at the expense of providing a less comprehensive curriculum. Being less comprehensive is not the goal, it is an unavoidable side effect of being less structured.

The downside of this has long been recognized. David K. Farkas, in his essay "Layering as a Safety Net of Minimalist Documentation" in *Minimalism Beyond the Nurnberg Funnel*[10, p. 249], noted:

> Minimalism is not without risks. Several commentators have expressed concern with reducing the information that users receive (Brockmann 1990, Farkas and Williams 1990, Redish, this volume, chapter 8). Carroll himself acknowledges a potential problem: "learners might not have access to enough information to reason successfully and might be anxious about bearing such responsibilities" (this volume, chapter 1). The risks, as I see them, are these:
>
> 1. The user may be unable to complete the task successfully.
> 2. The user may complete the task but expend more time and energy than he or she wanted to.
> 3. In the process of completing the task (or attempting to), the user may develop an incorrect mental model of the system that will cause difficulties later.

On paper, a comprehensive content set necessarily means a big book or a big set of books. And books, being paper, need to be organized top-down because bottom-up organization and navigation is not well supported on paper. You can't get more content without also getting more structure.

On the Web, though, this is not the case. As we will see in Chapter 22, a documentation set on the Web can look small on the outside while being large, comprehensive, and highly navigable on the inside.

Bottom-up navigation means readers are free to roam. An overt training structure is never imposed on them, and their desire to make their own curriculum is supported.

Bottom-up navigation means readers are free to roam.

Farkas makes the same case:

> We can ... use the rhetorical technique of layering to give users access to extra information that they may have need if minimalist documentation does not provide all the information they require....
>
> [T]he online medium had become dominant and ... the dynamic nature of online text and graphics is more suitable than is print to layering.
>
> —*Beyond the Nurnberg Funnel*[10, p. 247]

On the Web, or even in an EPPO help system, you can let learners explore for themselves without paying the price of being less comprehensive. And you can ensure that wherever their exploration takes them, learners will have content there to support them when they need it.

CHAPTER 18

Structured Writing

In Chapter 9, I said a topic type defines three things: the content, the order, and the form of a topic. That is what structured writing is all about: capturing, guiding, and validating the content, order, and form of a piece of content. The rules that describe the content, order, and form of topics can be written down in a content plan and audited and enforced by editors, or they can be encoded in data structures and validated by computer programs. Either way, the goal is to achieve greater consistency and reliability in how content is created and used.

Many writers have reservations about structured writing. They fear it will saddle them with pointless rules and structures that do not fit their writing. Structured writing, though, does not have to mean that everything is dictated from above. Indeed, one hallmark of professionals is that they govern their own work by defining structure for themselves. Professionals, after all, are not dilettantes or amateurs. Their ultimate satisfaction lies not in messing around with tools or with words (much as they may delight in them), but in providing a professional product that meets their customers' needs. True professionals are not self-indulgent, nor are they self-deluding. They know they are only human and that if they do not govern, discipline, and test their work against consistent, well-founded structures, they won't produce work of consistent quality and utility.

Structured writing is not the enemy of professional writers, but a natural and proper part of their tool chest.

Therefore, structured writing is not the enemy of professional writers, but a natural and proper part of their professional tool chest. And for occasional writers – those whose main job is something else but who are sometimes called on to produce content – structured writing can be a godsend if implemented properly. It guides authors and lets them know what is required and when they have completed the task in a satisfactory manner.

This is not to say that every writer's experience with structured writing has been a happy one. As with any other methodology, if it is not implemented properly it can be worse than the unstructured process it replaces. And the sad truth is that structured writing has often been implemented badly. However, the demands of the modern economy simply won't permit us to continue indefinitely with unstructured content and unstructured content processes. Whatever the disasters of the past, we have to master structured writing.

The demands of the modern economy simply won't permit us to indefinitely continue with unstructured content and processes

No matter what your role or responsibility, if you are a professional writer in industry today, structured writing skills are just as fundamental to your career as publishing skill were in the 80s and 90s.

The varieties of structured writing

There are many ways to specify the content, order, and form of information, and thus many forms of structured writing. If you have had any experience with structured writing or have read a book about it, you may have come away with the impression that it is one specific set of rules, one specific tool, one specific standard, or one specific XML schema such as DocBook or DITA. Nothing could be further from the truth. Structured writing is a wide and diverse field encompassing many methods, practices, tools, and technologies.

First, we need to distinguish two distinct but related types of structured writing, which I will call *rhetorically structured* writing and *computably structured* writing.

Rhetorically structured writing

I will use the term *rhetorically structured* to mean systems and approaches that formally define how the content, order, and form of information are expressed in order to make that information easier to consume and understand. Examples of rhetorical structure include the following:

- the traditional newspaper pyramid structure
- the classic essay structure of introduction, body, and conclusion
- the standard form of a recipe
- the standard form of an API reference
- Information Mapping, which sees an effective document as consisting of a collection of information blocks of defined types

The principle that an Every Page is Page One topic should begin by establishing its context is also an example of rhetorical structure.

Rhetorical structure includes rhetorical devices that apply broadly to many types of writing. For instance, Information Mapping defines six types of information which it claims can be combined into "information maps" (documents) that cover every kind of content on every imaginable subject. However, rhetorically structured writing also includes structures that are more specific to particular types of content such as those we looked at in Chapter 9. The classic structure of a recipe is a rhetorical structure designed to make it easier to understand how to prepare a particular dish, but it is a rhetorical structure specific to recipes. You would not use the same structure for information on fueling an ICBM or knitting a pot holder (Figure 18.1), even though these are all tasks.

The eye-catching design of the Checkered Pot Holders stands out in any kitchen.

Free Checkered Pot Holder Knitting Pattern

The perfect gift for your favorite chef! These checkerboard pot holders are knitted using the intarsia technique and then felted in the washing machine to make them good and sturdy.

Size
Before felting: Approximately 11x131/2 inches (28x34cm)

After felting: Approximately 8x81/2 inches (20x21.5cm)

What You'll Need

Yarn: 100% wool worsted weight yarn, about 121 yards (111m) each in 2 colors (makes 2 pot holders)

We used: Knit Picks Wool of the Andes (100% wool, 110 yards [101m] per 50g skein): #23420 Coal (color A), 2 skeins, #23775 Fog (color B), 2 skeins

Needles: US size 8 (5mm) straight, US size 9 (5.5mm) double-pointed, set of 2

Notion: Tapestry needle

Gauge
18 stitches and 24 rows=4" (10cm) in stockinette stitch before felting

Note: When changing colors, drop the old color and bring the new color up from under the old color, twisting them together to avoid gaps. For each section being worked, you will need 2 balls color A and 2 balls color B.

Making the Pot Holder
With color A and straight needles, loosely cast on 48 stitches.

Rows 1 and 2: Knit.

Row 3 (right side): With first ball of color A, knit 13, drop color A, attach first ball of color B, knit 11, drop color B,

Figure 18.1 – Knitting pattern rhetorical structure[1]

[1] the Editors of Publications International, Ltd. "Free Knitting Patterns for Beginners." 16 May 2007. HowStuffWorks.com. http://tlc.howstuffworks.com/home/free-knitting-patterns-for-beginners.htm 10 June 2013.

You don't need any special tools or techniques to conform your content to a rhetorical structure. You can write any of Information Mapping's six block types using a notepad and pencil. You can also create more specific rhetorical types such as recipes, knitting patterns, or API references in Word, Excel, or PowerPoint if you like. As long as you follow a consistent rhetorical template, you are doing structured writing – rhetorically structured writing – and that's a good thing.

Computably structured writing

I will use the term *computably structured* to describe systems in which content is encoded in a machine-readable format so it can be processed in multiple ways after it is written. Computable structures also deal with the content, order, and form of information, but they vary greatly in how strictly or loosely they define those things. Some implement very simple rules like every section must begin with a title, while others implement the rhetorical structure of specific information types.

Unless you write longhand on paper, type on a typewriter, or carve stone tablets, your content is being captured and stored in a computably structured format. The structure and semantics of that format determine what you can do with your content.

Strictly speaking, all data formats are computably structured. All data in a computer system is structured in some way, or the computer can't deal with it. But we usually reserve the term *structured writing* for formats that are defined separately from the applications that create or consume them.

An application like Word or PowerPoint is not designed to share data with other applications (except its Office stable mates). You usually choose this type of application for its specific list of features and functions. If you choose an application for its functionality, you usually aren't interested in the structure of its file format, and, therefore, you are not doing structured writing in the computably structured sense (though you may be doing something highly structured in the rhetorical sense). You don't adopt the Word file format for any property of the format; you adopt it because you bought Word.

On the other hand, if you begin by choosing (or designing) a file format and then go shopping for tools that can create or process that file format, you are doing structured writing in the computably structured sense (though you may be doing something completely unstructured in the rhetorical sense). If you decide to adopt DocBook or DITA or develop your own XML schema or database structure, you generally make that decision first and then go looking for tools.

By this definition, using XML does not automatically mean you are doing structured writing. For example, XML is used to encode the files used by many closed applications, including Microsoft Word. In fact, a closed application could use DocBook or DITA as an internal format under the covers, and you still might not be doing structured writing if you aren't conscious of the structure and have no plans for the data file beyond what the application does with it.

Using XML does not automatically mean you are doing structured writing.

Essentially, you are doing computably structured writing if you are aware of the structure, creating it deliberately, and specifying how it will be processed. You may use high-level tools and you may even hide much of the structure from authors, but taking ownership of the data structure and how it is processed puts you in the computably structured writing camp.

You are doing structured writing if you are aware of the structure, creating it deliberately, and specifying how it will be processed.

In technical communication, the most common way to create a data structure for content is with XML. Technical communicators use standard XML schemas like DocBook, DITA, or S1000D; industry vertical schemas; or custom schemas.

On the Web, in the WCMS (Web Content Management System) world, the most common way to create a computable structure is with a relational database. Standard WCMS platforms such as WordPress and Drupal come with standard database schemas for content, and WCMS developers frequently create custom schemas for particular content types. Technical communicators generally use CMSs or CCMSs (Component Content Management Systems) to manage chunks of content structured in XML, but they don't use the underlying database to model content structure. Other structured writing systems use a software version control system or a plain file system to store their XML files. Some fusion of the technical communication and Web approaches is now being attempted. At time of writing, it is a little too early to tell where those attempts will lead.

The data structures that you choose may or may not support or enforce a rhetorical structure. For instance, DocBook supports some of the structures commonly found in technical documents, but it does not constrain how you use those structures. You can write a computably structured document in DocBook that has no rhetorical structure at all. You can still call that structured writing – computably structured writing – but it really has nothing to do with rhetorically structured writing.

A word about SPFE

I am working on a project I call the SPFE[2] architecture, which is a structured writing architecture designed specifically to support the creation and management of EPPO topics. Strictly speaking, SPFE is not a competitor to DITA or DocBook. In fact, you could use either DITA or DocBook as a layer within a SPFE system. SPFE isn't a schema either, but an architecture designed to make it inexpensive to develop custom topic type schemas for your own content. The SPFE architecture is also designed to natively support bottom-up organization and the soft-linking methodology I describe in Chapter 20. A SPFE Open Toolkit is in development. If you are interested, you can follow its progress at SPFE.info and SPFEOpenToolkit.org.

Other forms of computable structure

Not all computably structured writing uses XML or a relational database. For example, the JavaDoc code documentation system extends the native Java comment syntax with additional markup. JavaDoc comments begin with `/**`, and can contain certain HTML markup tags as well as some special tags, including: `@author`, `@version`, `@param`, `@return`, and `@exception`. The special tags introduce sections that should appear in every JavaDoc API entry. In other words, these tags define the rhetorical structure of the API reference entry, and they do it in a way that is computable. This allows a JavaDoc processor to organize, format, and link a Java API reference document based on function signatures and code comments.

This too is structured writing. In forming your structured writing strategy, it is important not to overlook existing sources of structured content just because they are

[2] The name SPFE is an acronym for the four layers of the SPFE architecture: Synthesis, Presentation, Formatting, and Encoding. It's pronounced "spiffy."

not in XML. Nor should you assume that if you want to get structured content from a particular group of contributors, you have to get them to write in XML. Any format that captures the content, order, and form you need in a computable manner will work, and, often, non-XML formats will be easier for contributors to learn and use.

Open and closed formats

Figure 18.2 shows some of the various forms of structured writing, both rhetorically structured and computably structured. I have broadly defined the types of structure into general, formal, and subject-specific.

	Closed Format	Rhetorical Structure	Open Format
General	MS Word FrameMaker	General Text	DocBook Markdown Wiki markup
Formal	PowerPoint WordPress	Information Mapping DITA*	DITA EPPO-Simple
Subject-specific	TurboTax Intwined Patern Studio	Recipes API References EPPO	MathML RecipeML Custom

Figure 18.2 – Structure matrix

I'm using these terms as follows (you won't necessarily find this usage elsewhere):

- **General:** General structure means that content is structured only in the general ways that all documents are. Books, articles, and web pages that don't have a more specific structure are at least recognizably structured as books, articles, or web pages. Closed tools that create generally structured content include Word and unstructured FrameMaker.[5] Open format languages that create generally structured content include DocBook,[6] HTML5, wiki markup, and Markdown.[7]

- **Formal:** Formal describes content that is structured more formally than general document structures. For example, Information Mapping is a formal system for composing documents from standard (formalized) content blocks. DITA provides a rhetorical structure, though it is also a data format. DITA provides three basic topic structures, concept, task, and reference, which provide a formal structure you can follow even if you do not use DITA XML markup to do so. In fact, the division of help systems into concepts, tasks, and reference is quite common.

  Closed applications that create formal structures include PowerPoint, which divides content up into fixed-format slides, and WordPress, which has a fixed format for blog posts. (The WordPress database format for blog posts isn't proprietary, but it is not intended to be used outside of WordPress, so it is closed for the purposes of this discussion.)

  Open formats that create formal structures include base DITA and EPPO-simple, which is a set of schema components that is (at time of writing) under development as part of the SPFE Open Toolkit. It is intended for creating subject-specific EPPO formats, but it could also be used alone as a formal-level EPPO format.

- **Subject-specific:** A subject-specific structure is used only for writing about a specific subject. The recipe format is subject specific. So is the API reference format. Wikipedia's ad hoc formats for cities, actors, politicians, etc., are subject-specific formats. If you follow a formula for writing about a specific subject, you are using a subject-specific rhetorical format. EPPO fits in this box because the EPPO design pattern calls for the use of subject specific topic types.

  It's hard to find examples of closed-format, subject-specific applications – though I was able to find a program called Intwined Pattern Studio[8] for turning knitting

---

[5] Of course, FrameMaker also supports open structured formats like DITA.

[6] DocBook does have some elements that are designed specifically for use in technical communication, particularly related to software, but the overall structure is general. This book was written in DocBook.

[7] Markdown is a simplified language for writing Web pages using markup inspired by the way people format plain text email. See http://daringfireball.net/projects/markdown/.

[8] http://intwinedstudio.com/

patterns into written instructions – but there are many that are more data oriented. TurboTax is an example of an application that is designed only for the subject of income tax returns.

There are a few standardized subject-specific data formats, such as MathML and RecipeML, and some industry-vertical formats, but most of the subject-specific data formats are custom made by the organizations that use them. Some are DITA specializations, some are custom XML schemas, and some are custom database schemas. EPPO-simple is intended to provide building blocks for subject-specific EPPO formats. Most subject-specific data formats include some general or formal structural elements for data that does not require subject-specific structures. Examples of this include base (unspecialized) DITA and EPPO-simple. Another example would be the use of Markdown in a Web Content Management System (WCMS) for fields that contain general description.

Any document you create on a computer will fit into one of the rhetorical structure classes and either the open or closed format. So, for instance, if you wrote a recipe in Word you could map it on the matrix as shown in Figure 18.3.

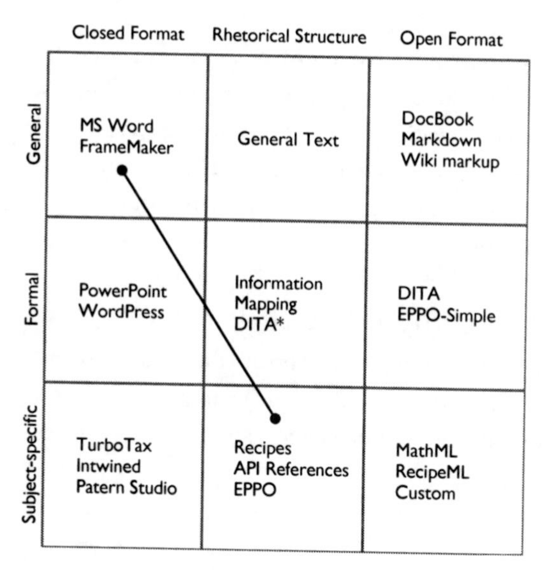

Figure 18.3 – Writing a recipe in Word

There are many options for writing EPPO topics, as you can see in Figure 18.4.

	Closed Format	Rhetorical Structure	Open Format
General	MS Word FrameMaker	General Text	DocBook Markdown Wiki markup
Formal	PowerPoint WordPress	Information Mapping DITA*	DITA EPPO-Simple
Subject-specific	TurboTax Intwined Patern Studio	Recipes API References EPPO	MathML RecipeML Custom

Figure 18.4 – EPPO authoring choices

EPPO is a design pattern, not a fixed format, so you can write EPPO topics in any authoring application. There are several good reasons to choose a structured-data approach, and subject-specific structured data formats in particular can provide excellent author guidance, validation support, and link automation (see Chapter 20), but EPPO does not require this approach. You can start writing EPPO topics today in whatever system you are using now.

The varieties of computable structures

Computably-structured writing systems break content up into separate parts and label the parts so that a computer can address each part and process it separately. Here are some examples of different writing systems and the processing each can support.

```
<movie-review>
  <title>Rio Lobo</title>
  <body>
     <p><director name="Howard Hawkes">Hawkes'</director> final
     film is a lighthearted Western in the <movie>Rio Bravo</movie>
     mold, with <actor name="John Wayne">the Duke</actor> as an
     ex-Union colonel out to settle some old scores. </p>
  </body>
</movie-review>
```

Figure 18.5 – Computably-structured content example

Figure 18.5 shows an example of computably structured content in XML. The XML tags divide the content into distinct elements so a program can address each separately. This is a subject-specific format for movie reviews. Its root element (the one that wraps everything else and establishes the document type) is `movie-review`. The document also contains other elements specific to the subject of movies, such as `director`, `actor`, and `movie`. This enables you to process the content in ways that are specific to movie reviews. For instance, suppose you had a collection of movie-related information such as reviews, actor bios, etc. A program could create a list of every movie reviewed in the collection by looking at the content of the `title` elements. Or by looking at the contents of the `actor` element, it could create a collection of all the movie reviews that mention John Wayne.

By contrast, Figure 18.6 shows the same text, but in a general format that is not specific to its subject. In this case, the root element is `html`, and there is nothing in the markup that is specific to the subject of movie reviews. There is no way a program could pull a list of the titles of all the reviewed movies in a collection based on this markup because the collection does not contain the necessary subject-specific markup.

```
<html>
   <h1>Rio Lobo</h1>
   <p>Hawkes' final film is a lighthearted Western in the Rio
Bravo mold, with <a href="http://johnwayne.com">the Duke</a> as an
ex-Union colonel out to settle some old scores. </p>
</html>
```

Figure 18.6 – HTML structured content

For yet another example, Figure 18.7 shows the same passage with DocBook markup.

```
<article xmlns="http://docbook.org/ns/docbook"
         xmlns:xl="http://www.w3.org/1999/xlink" version="5.0">
  <title>Rio Lobo</title>
      <para>Hawkes' final film is a lighthearted Western in the
          <citetitle>Rio Bravo</citetitle> mold, with
          <link xl:href="http://johnwayne.com">the Duke</link>
          as an ex-Union colonel out to settle some old scores.
      </para>
</article>
```

Figure 18.7 – DocBook structured content

DocBook is a little more specific about the document structure. It identifies the root element, `article`, and offers additional semantic elements, like `citetitle`. However, this is still a general structure, unrelated to the subject matter. DocBook's structure is even a little more general than HTML markup. HTML is specific to the web, whereas this markup is designed for any kind of article for any kind of media.

The capabilities of HTML have been considerably expanded with HTML5. HTML5 adds some basic document structured markup to HTML, including tags like `article` and `section`, bringing it slightly closer to what DocBook provides, though DocBook remains far richer in document semantics. HTML5 also supports microformats and microdata, which allow some subject semantics to be added to an HTML document. This can be useful to consumers of HTML5 content, but these systems provide semantic annotation of general structures rather than enforcing rhetorical structures. HTML5 is therefore not a likely candidate for a structured authoring format, but it is certainly an output format that a structured authoring system should support.

Benefits of computably structured writing

Structured writing, especially computably structured writing, provides a variety of benefits, some of which depend on the format. Each benefit described below is attainable with the right format, but not all formats provide all of these benefits. Also, the benefits are not specific to EPPO topics, but apply to any information design pattern.

Improved content quality

The first and most important reason to adopt structured writing is to improve the quality of your content. Defining firm rhetorical structures for all your content can help ensure that content is complete, consistent, and navigable. Using computably structured writing tools to capture your rhetorical structures improves quality further by giving writers better guidance as they write.

Guidance for writers

Writing remains a craft. It relies on the experienced touch of the individual writer to say the right thing and to say it well. But like professionals in other crafts, writers can improve the quality and consistency of their work by using guides and templates. Well-designed, rhetorically and computably structured schemas provide a guide and template that can ensure greater consistency and accuracy while enabling writers to work more quickly and with greater confidence.

Conformance and quality

One key characteristic of an EPPO topic is that it conforms to a type. You can use computably structured writing to help ensure that your content conforms to its type.

When people make the case for investment in technical communication, they often point to technical communication problems that have contributed to notable accidents. Using a schema that requires each component to be present helps prevent errors and omissions that could lead to user frustration (at least) or catastrophic loss (at worst).

By enforcing your schema in your authoring tools, you give writers and editors direct, immediate feedback, which can help them become more productive, both in terms of the speed and the quality of their work.

Linking

Every Page is Page One topics link richly along lines of subject affinity. You can use structured markup to capture subject affinities in your content and use them to generate linking automatically. See Chapter 20 for more information.

Content manipulation

Computably structured content lets you treat your content like a database and write query expressions against it. Thus you could run queries such as: show me all the movie reviews that mention both John Wayne and Howard Hawkes, or give me a list of all the API routines that take or return a `config record` data structure. If you were writing a topic about the `config record` data structure, that query could be used to insert a list of relevant API routines into that topic, and unlike a list that was compiled by hand, it would automatically be updated if the API changed.

The possibilities for useful content manipulation are broad. The more subject-specific your data format is, the more you will be able to manipulate your content.

Future proofing

Working in an open format can help ensure that your content will be accessible and usable in the future. This is not simply a matter of selecting a particular data format. Just because a format is open does not mean it is going to be around for ever. In fact, open data formats are just as vulnerable to changes in technology and market conditions as proprietary ones. A good example is XML's predecessor, SGML, which was adopted by many as a future-proof format but has now largely fallen out of use.

However, an open format falling out of use does not mean the content is lost. (It does not mean it when a closed format does either, as newer applications will often read the older formats to encourage migration.) You can always translate your content from an older format to a newer one. The real question is, will the content have the structure you will need in the future?

What you really should care about for future proofing are the following two things:

- **Semantics:** You need to preserve the semantics of your content in a usable form. For example, you can translate your old WordPerfect files to DITA, but the translation will be imperfect because DITA demands semantic information, such as clearly delineated procedural markup, that WordPerfect files don't contain. Therefore, you will need to do some cleanup, which will cost you time and money. In general, the more semantic information your content contains, the easier it will be to translate it into a future format.

- **Medium:** The nature of your current output medium will affect how easy your content is to work with in the future. No matter what file format you use, books don't translate well to the Web. You can't translate long and linear to short and cross-linked. Ultimately, the media your content was written for and the way it is organized is a far bigger hurdle to future use of the content than the file format. Even if you are not delivering to the Web today, creating EPPO topics rather than writing linear books is the best way to future proof your information.

Single sourcing

One of the most common phrases used to describe structured writing is "separation of content from format." This refers to the ability to create content once, in a format-independent way, and then generate output to various media by adding formatting appropriate to each medium. That is, the content is separated from the specifics of fonts, margins, etc., and expressed in terms of the generic structures of a document such as headings or tables. This can be achieved using a general-structured format such as DocBook (see Figure 18.7).

However, separating content from formatting is not enough to enable you to organize content differently for different media. More than formatting separates the linear organization of a book from the random-access structure of the Web. Separating content from formatting is not enough to change top-down organization to bottom-up organization.

It is also worth noting that any word processor or desktop publishing program with stylesheet capability is capable of separating content from formatting. You don't need XML or a relational database to do this.

Reuse

While single sourcing focuses on outputting one publication to multiple formats, reuse looks at outputting one piece of content to multiple publications or displaying it in multiple contexts. Reuse generally depends on small units of content that are authored with reuse in mind. This can be done with formal systems, like DITA's concept, task, and reference topics, or subject-specific systems (see Chapter 21 for more on reuse).

Content exchange

If you want to exchange content with other people, you need to deliver it in a format they can process. The simplest way to achieve this is to have both the sender and the recipient use the same format. If both parties use Word, FrameMaker, DocBook, or standard DITA, exchanging content is relatively simple.

The problem is, not everyone wants to receive content in the same format. So, at least in some cases, you will have to translate your content to the format the receiver is expecting. As with future proofing, the key to this is not writing or storing content in the same syntax you want to deliver – it is easy to transform content from one syntax to another.[9] The real key lies in the semantics. You can always exchange content (or data) if the other party's data has the same semantics as yours (regardless of syntax) or if your content is semantically *up hill* from their content. That is, you can exchange content if the semantics of your content can be transformed into the semantics of their content without loss. This can happen if your content semantics are richer than theirs, but not if they are poorer.

Consider the movie-review content examples in Figure 18.5 and Figure 18.7. The markup that is specific to movie reviews could easily be translated into DocBook markup for delivery to someone who uses DocBook or to HTML5 with microformats for a content API, but the DocBook or HTML5 version could not be translated as easily or reliably into the movie-review markup. The movie-review markup is semantically up hill from the DocBook or HTML5 markup.

Of course, creating subject-specific markup has its costs as well, so you have to weigh the costs and benefits. But the conformance and quality advantages are compelling, especially for Every Page is Page One topics, which work best when written to a well-defined structure. And because you can easily translate to a more general format like DocBook, you can take advantage of all the publishing tools available for DocBook.

---

[9] If you are creating a *content API*, you can give your API the capability of delivering content in more than one format.

If you use DITA, you can use the DITA specialization mechanism to create subject specific markup that will retain valuable content semantics that will be useful even if you move away from DITA in the future.

Structured writing and bottom-up organization

We saw in Chapter 5 how important bottom-up organization and bottom-up navigation is in a content set that people often access via search or links. We also saw that the main challenge in creating bottom-up organization and navigation is that the bottom-up organization of a web can't be usefully viewed from the top down.

The problem is, how do you plan, execute, monitor, and validate an organization and a navigation schema that you can't view from the top down? Manual forms of organization tend to fall apart because we depend so much on the ability to step back and see the big picture. But with a web, the big picture is just a jumble of intersecting relationships (Figure 5.2, "Map of a complex web"). Bottom-up organization is only visible and comprehensible from the bottom up.

This does not mean manual organization is impossible. Wikipedia makes it work with thousands of contributors and a cadre of supervising editors. However, each contributor creates the content and the organization from the bottom up.

Some of the most valuable relationships between topics don't fit into neat taxonomies or hierarchies. Instead, they arise from the same particular and irregular relationships that exist in the real world. These irregular relationships are often the most valuable ones in the content set because they address those situations where the task or the product does not fit a conventional mental model and, therefore, is more likely to give the user trouble. But these irregular subject affinities are also difficult to see and manage from the top down.

While bottom-up organization and navigation and irregular subject affinities can be difficult to manage by hand, they can be managed effectively by an algorithm. But for the algorithm to work, it needs computably structured content. Computably structured writing can be a big help in organizing content from the bottom up.

CHAPTER 19

Metadata

Metadata plays a variety of roles in the creation and management of Every Page is Page One topics. But if *structured writing* is a term that causes confusion, *metadata* is doubly so. Metadata is used in many ways for many purposes, but often people see metadata in just one role – as the information attached to a document or a web page to help people find it. This use of metadata is essentially the same as the label on a product package, such as that in Figure 19.1.

If you prepare a document using tools on the Closed Format side of the content structure matrix (Figure 18.2, "Structure matrix") and then submit that document to a WCMS, you will probably have to fill out a metadata record for the file as part of the submission process. The WCMS will then use some or all of that metadata to build navigation aids on the website. The WCMS puts a label on your document similar to the label on the jar of pasta sauce – that is, a label that describes the contents of the document.

Figure 19.1 – Pasta jar label

But if you use an authoring method from the Open Format side of the matrix, you will be creating a lot more metadata, and most, and possibly all, of the metadata will be created before or during the authoring of the content, not afterwards. For example, the movie-specific markup in Figure 18.5 contains much of the identifying metadata, suitably labeled, in the topic markup itself. Even if you are not doing structured writing, though, metadata can play just as important a role in the authoring process when you are creating content that has a formal or subject-specific rhetorical structure. To put it simply structure is defined by metadata. But the metadata that defines structure is not on the outside of the content, like the label on a pasta jar, it is inside the content, labeling its individual parts and pieces.

The meaning of metadata

Metadata is simply information about data or, to put it another way, data that describes other data. The label on the jar in Figure 19.1 is just data because the stuff in the jar isn't data, it's pasta sauce. The label on a document in a WCMS is metadata because the document is itself data.

Metadata is ubiquitous. Indeed, most data is useless without metadata to tell us what it means. And because metadata is also data, we need metadata to tell us what the metadata means.

Take, for example, an XML document. An XML document contains markup, which is a form of metadata. Consider this fragment:

```
<p>The <library-name>foobar</library-name> library contains
   the routines <routine-name>foo()</routine-name>
   and <routine-name>bar()</routine-name>.
</p>
```

All the bold bits are metadata. The `<p>` tag is metadata that tells us that the string of characters between it and the closing `</p>` tag is a paragraph. The `<library-name>` and `<routine-name>` tags are metadata that tell us that the strings they delimit are library names and routine names, respectively.

The set of tags allowed in an XML document is defined in a schema. A schema that describes the tagging used in the above fragment might contain lines like these:

```
<xs:element name="library-name" type="xs:string"/>
<xs:element name="routine-name" type="xs:string"/>
```

These lines are metadata that describe the `<library-name>` and `<routine-name>` elements. They say, for instance, that a `<routine-name>` element can't contain any other elements, only character string data. So, a schema is metadata that describes a tagging language, and a tagging language is metadata that describes content.

Tables of contents and indexes are metadata. So are headings, subheadings, captions, and running headers and footers. These things are all data that describe the content of the book (which is data), and they are therefore metadata.

The same sort of thing applies in other kinds of data. You will find layers of nested metadata wherever you look. In the database world, we have column names, which are metadata, and data dictionaries, which are metadata that describes the columns and the relationships between them.

Not all forms of metadata are referred to as *metadata*. Many forms of metadata have their own long-standing names: index, schema, data dictionary, table of contents, catalog, tag, label, and so forth. But on the Web and in content management systems, the proliferation of new forms of metadata and new ways of capturing and using metadata seem to have have outstripped our ability to name them all. Or perhaps many of them already had names, but people felt the connotations of the old names (such as index) might obscure what was needed in the new environment.

Whatever the reason, people started to use the generic term *metadata* for new kinds/forms/presentations of metadata. So today we have many forms of metadata that have their own names and many other forms, often closely analogous to the existing forms, that all go under the generic moniker *metadata*. Thus the confusion: metadata can refer to a bunch of things that are all called metadata and to a bunch more things that are not commonly called metadata, but which still are metadata.

Topics should merit their metadata

In *Everything Is Miscellaneous*[28, p. 105], David Weinberger argues that the secret to making information findable is not organization, but metadata. In other words, as we saw in Chapter 3, web pages are organized dynamically by the Web itself, and the thing that makes it possible for the Web to filter content accurately is metadata (both the metadata authors create while writing the content and the metadata readers create by tagging, liking, or linking to the content). Without explicit metadata, search engines will attempt to derive metadata from the content itself. The Web is driven by metadata.

It is not hard to add metadata, but it is hard to add good metadata. If your content is going to get filtered in when and where it should be, its metadata has to accurately reflect the topic. If your metadata is not correct, it will foul the filter, and your content will get punished.

The Web is driven by metadata.

The problem is, you can't arbitrarily attach metadata to an object that does not deserve it. You cannot attach more metadata to an object than that object's intrinsic properties deserve, at least, not if you want a reliable result. If the pieces you are labeling are too small or if their boundaries are poorly defined, the metadata will not fit properly.

Think about the label on a carton of ice cream. For flavors that contain nuts, there should be a warning on the container that says "contains nuts." That is the correct scale for that label. Attaching the label "contains nuts" to the entire ice cream shop is technically correct, but it would unnecessarily exclude someone with a nut allergy from eating any flavor of ice cream, even strawberry. At the other end of the scale, you can attach the label "contains nuts" to the nuts individually, but then customers with a nut allergy might order a flavor that contains nuts and not discover that it contains nuts until they get to the nuts themselves. The most useful scale for labeling ice cream is not the store or the individual ingredient, but the carton or serving size.

The same is true of content. You want the metadata to correctly label the serving size of content the reader will be consuming. But with books burst into topics, or built up from reusable building block fragments, that is not always easy. Tom Johnson wrote about this in his blog post "The Importance of Chunking for Sorting." He acknowledges that chunking content too finely can cause problems when you try to retrieve that content with a query:

> [I]f you pull together all topics that have specific metadata, such as all topics related to scheduling events, you may get an unordered collage of topics. The order of the topics may not reflect any kind of sequenced or arranged reading. The list of topics no longer forms a larger, well-written chapter that contextualizes each topic, but rather may seem like little scattered objects here and there.
>
> —Tom Johnson, "The Importance of Chunking for Sorting."[1]

The problem Johnson describes occurs because all those finely chunked pieces no longer merit the metadata that is attached to them. Consider what happens when you take apart a piece of machinery such as an old-fashioned alarm clock.

[1] http://idratherbewriting.com/2011/04/18/the-importance-of-chunking-for-sorting/

We can attach several pieces of useful metadata to the clock itself. It is a device for telling time. It is a device for waking you up. It is (possibly) an item of home decor. It is (potentially) a movie prop for a period piece. There are a number of useful properties that you could assign to it that would help you find it when you need something to fulfill one of those functions.

Now start taking it apart. First, disconnect various assemblies: the case, the clock mechanism, the ringer. Some of these could still have interesting metadata attached to them. But as you continue taking each of these assemblies apart, you are left with a collection of screws, gears, and pieces of bent metal. At this point, it is no longer useful to attach metadata to them that says they used to be part of an alarm clock.

Figure 19.2 – Clock parts[2]

The screws and gears in Figure 19.2 could have other metadata attached to them, metadata that describes them as individual pieces. As individual pieces they have definite characteristics, such as thread count, circumference, or number and spacing of teeth. None of this is metadata for the alarm clock, and none of the metadata that applied to the alarm clock properly belongs to the individual gears and screws. They are potentially reusable for all sorts of other projects that may have nothing to do with the alarm clock. Putting alarm clock metadata on the gears and screws would

[2] Photo by Vassil, Wikimedia Commons, Public domain

only obscure their proper screw and gear metadata, potentially hindering their reuse for other purposes.

Suppose you do attach alarm clock metadata to these parts. Then if you query that metadata, what you will get back is not an alarm clock, but a pile of gears, screws, and bits of bent metal. That is not an alarm clock, it does not serve any of the purposes of an alarm clock, and it does not merit alarm clock metadata. Such metadata is not accurate or reliable. It is not useful to anyone who needs an alarm clock.

This does not apply only to the metadata attached to the whole topic. It applies to its internal metadata as well. Once gears and cogs are assembled into a clock they become parts of a clock in a way they were not when they were sitting in the parts bin. An item of content like "6 eggs" is not an ingredient of an omelet until it is included in an omelet recipe. By itself it might just as well be part of a grocery list or a math problem. But once it is included in a recipe, it becomes an ingredient of that recipe and can be reliably identified as such.

Useful metadata can only be applied to useful things. Useful content metadata can only be applied to useful units of content. Break content down into chunks that are smaller than is useful to a reader and you cannot attach metadata to it that will be useful to the reader.

If content has been aggressively decomposed into small fragments to maximize reuse or optimize translation memory, those fragments will generally not merit EPPO topic metadata. They should have their own metadata, just as the springs and gears of the disassembled clock have their own metadata, which is not clock metadata. And when you assemble those fragments into EPPO topics, you need to give those topics their own topic metadata.

So, if you want your topics to have metadata that will help your readers find and organize your content, you need to create topics that actually merit a range of useful metadata. Every Page is Page One is about creating single-serving helpings of information that merit rich metadata that makes that information easy to find and satisfying when found.

The qualities a piece of useful content must have to merit its metadata turn out to be the same qualities that EPPO topics have.

- **Self-contained:** To be labeled accurately, an object must be self-contained. If it is a component of something larger, the topic label belongs on the larger object.
- **Specific and limited purpose:** Metadata is essentially a description of what a piece of data does. If the purpose of the content is not specific, there is no way to attach specific metadata to it. If the purpose is not limited, the metadata required cannot be limited either, and unlimited metadata is as bad as no metadata.
- **Conform to type:** A well-defined content type defines every aspect of a piece of content, and metadata labels each aspect of the content. If the aspects of the content are not consistent, the metadata cannot be consistent. And inconsistent metadata is of little use. If you can't rely on the metadata, you can't use it to find, create, manage, or maintain content.
- **Establish context:** When a topic explicitly establishes its context for the reader, it confirms that the metadata for the topic is correct. Showing the metadata as part of the topic, as we saw in the Blue-footed Booby topic in *All About Birds* (see Figure 10.4), is one of the most useful ways to establish context.
- **Assume the reader is qualified:** Every topic is written for someone. The metadata should identify the audience, either explicitly or by implication. If the topic does not stick to what is stated or implied, it does not merit the metadata.
- **Stay on one level:** As with reader qualification, the metadata should identify the level, either explicitly or implicitly. If the topic does not stick to that level, it does not merit the metadata.

If you want to create EPPO topics out of smaller building blocks, that's fine. But you need to recognize that just as the metadata merited by individual parts in the parts bin is not the same as the metadata merited by a working clock, so the metadata that is appropriate for labeling your reusable content chunks is not the metadata that applies to an EPPO topic built from them.

Metadata comes first

To create content that truly merits its metadata, the best thing to do is to start with the metadata first. That is, define the metadata up front – all of it – and then write a topic that merits that metadata.

Create metadata first and content afterwards.

In fact, this would make an excellent definition of structured writing, whether we are talking about rhetorically structured, computably structured, or both: when you do structured writing, create the metadata first and the content afterwards.

This approach is similar to the test-driven development[3] practice that is part of agile software development. In test-driven development, you don't start by coding a routine and then figure out how to test it; you write the test first and then write code that passes that test. Writing the test first is an excellent way to make sure you really understand what the code is supposed to do. This has been shown to produce higher quality software with fewer bugs.

In many ways, the metadata for a topic is a test for the topic's content. It defines the purpose, type, and level of the content in a way that makes it clear to the writer, editor, and reviewers exactly what is required. If you use computably structured writing, you can also use the metadata to audit individual topics and the topic collection.

The first and most important piece of metadata about an EPPO topic is its type. The most elementary part of defining the metadata first is to choose the topic type before you start writing. If you are doing computably structured writing, this means choosing the appropriate schema up front. If your topic type calls for additional metadata, such as classification or index terms, write those first. This will help you make sure you know exactly what content you are supposed to create.

[3] http://en.wikipedia.org/wiki/Test-driven_development

CHAPTER 20

Linking

Linking is an important aspect of Every Page is Page One topic design. Unfortunately, many conventional tools make linking expensive to create and manage. I won't go through the linking mechanisms of all the available tools – they are adequately documented elsewhere. Instead, I am going to describe two alternate methods for linking: crowdsourced links and soft linking.

Crowdsourced links

Crowdsourced links are generated by the community rather than by authors and editors. Wikipedia articles tend to be richly linked because Wikipedia has thousands of contributors constantly adding to and maintaining topics. If you are using a Wiki and have a fairly open access policy, you may be able to crowdsource some of your linking. Social platforms also offer the ability to crowdsource linking through comments, and you can ask users to suggest related resources in comments.

Crowdsourcing may also be possible with other tool sets. The main challenge is to get people involved in linking content. The key is to make people aware of the value of links and the importance of a robust bottom-up content organization. I would love to hear from you if you have made crowdsourced linking work in your organization.

Soft linking based on subject affinities

Soft linking is a technique based on subject affinities. It is not supported by default in most tools, though you may find add-ons for some tools that support this technique or something similar. You will also find that many database-driven content sets are effectively using this technique, building links based on queries of the database content. Soft linking is a technique in which links are generated automatically based on subject affinities recorded in structured data. Soft linking is not difficult to implement. The secret is to have content that properly identifies and fulfills its purpose so the system can form useful links to the right content.

Example 20.1 shows a passage from a movie review with the significant subjects it mentions – its subject affinities – marked up.

Example 20.1 – Subject affinity markup

```
<p><director name="Howard Hawkes">Hawkes'</director>  final film
   is a lighthearted Western in the <movie>Rio Bravo</movie>
   mold, with <actor name="John Wayne">the Duke</actor> as an
   ex-union colonel out to settle some old scores.
</p>
```

The subject affinities in this passage – the important things that it mentions – are the director, Howard Hawkes, the actor, John Wayne, and the movie, Rio Bravo. There are other affinities worth marking up, such as the genre, but for simplicity's sake, I have just marked up those three.

As we saw in Chapter 5, links in Every Page is Page One topics are not usually used to make explicit references to particular documents. They are mostly used to provide access to related content along lines of subject affinity. With soft linking, authors don't need to identify a particular resource as the link destination. All they need to do is to mark up the subject affinities. Scripts can then be used to find topics on related subjects, such as a biography of John Wayne or reviews of Rio Bravo.

Here are some advantages to not requiring authors to identify particular resources:

- It saves time. If all authors have to do is note that a particular string has a subject affinity, they can work much more quickly than if they have to stop and identify a resource to link to. The longer it takes authors to create links, the fewer links they will create.
- While you're developing an information set, the resources you might want to create links to may not exist yet. But even if you can't create a hard link to a resource that does not exist, you can note an affinity to the subject.
- Relying on authors to find resources to link to can result in inconsistent linking. Different authors may have different ideas about which resources are the best to link to, and authors can only link to what they can find, so they may end up linking to less apt resources they already know about rather than to better resources

they don't know about. Author behavior, like that of readers, tends to be satisficing. Authors generally won't keep looking for the best link target once they have one that seems good enough. They will pick the first "good enough" target they find, resulting in a topic set full of "good enough," rather than excellent, links. The larger the set of topics you are dealing with, the longer it will take to find the perfect link target and the more likely it is you will find a lot of inferior targets, which will drive down the quality and quantity of your links. Even with a CMS, the cost of links still increases as the number of topics increases.

- If you reuse content in different information products, you may find that trying to reuse a topic that contains hard-coded links will result in broken links because the target topics don't get included in the new information set. One way to handle this is through indirection – using a map file to point to different link targets for each information set – but that creates a lot of overhead. If you simply note the subject affinities, the soft links can be resolved at build time to point to available resources in the current information set.

Soft linking based on subject affinities does not require XML. Any data format that allows you to highlight subject affinities in a computable manner can be used.

If you can identify subject affinities through natural language processing or structured pattern matching, you may not need to have authors highlight subject affinities at all. I have used this technique in several projects to find API routine names in code samples and generate links from each of them to the API reference.

Soft linking is not indirection

What follows is the most technical thing in this whole book. I include it because it is important not to confuse soft linking with another technique known as *indirection*. Both techniques are useful – and each has pluses and minuses – but they are different, and you can miss the true power of soft linking if you confuse it with indirection.

The simplest way to distinguish indirection and soft linking is to look at the text. When you use indirection, you create a link in the text, when you use soft linking, you don't. Let's look at examples of direct, indirect, and soft linking.

First, here is a direct link:

```
<p>Hawkes' final film is a lighthearted Western in the Rio Bravo
   mold, with <a href="http://johnwayne.com">the Duke</a> as an
   ex-Union colonel out to settle some old scores.
</p>
```

This is the familiar HTML `<a>` element with an `href` attribute that contains the direct address of the resource to link too. It is thus a direct link. Here is an indirect link:

```
<p>Hawkes' final film is a lighthearted Western in the Rio Bravo
   mold, with <link idref="john-wayne">the Duke</link> as an
   ex-Union colonel out to settle some old scores.
</p>
```

With an indirect link, the `<link>` element contains an `idref` attribute with a link ID. That link ID is then listed in a linking table like this:

```
<link id="john-wayne" href="http://johnwayne.com"/>
```

The actual destination of the link is stored in the linking table rather than in the source document. This has a couple of advantages. First, you can change the destination of a link by updating the linking table rather than having to update the source document. Second, if you reuse the source document in more than one place and want to have it link to different content, you can write a linking table for each place.

Indirection is a useful tool, but it is not the same thing as soft linking. Here is the soft-linking example again:

```
<p><director name="Howard Hawkes">Hawkes'</director> final film
   is a lighthearted Western in the <movie>Rio Bravo</movie>
   mold, with <actor name="John Wayne">the Duke</actor> as an
   ex-Union colonel out to settle some old scores. </p>
```

There is no link in the text. Instead, the semantics (meaning) of the source text has been clarified. The build system knows that the `<actor>` tag identifies an actor's name. That is, it highlights a subject affinity. To create a link on the text of that element, "the Duke," the build system runs a query looking for topics or pages on John Wayne. It can look for these anywhere, in any source. It doesn't matter how those resources identify themselves or what schema they are encoded in. All that matters is that the build system knows how to retrieve relevant resources.

In direct linking, you have a pointer from one topic to another. In indirect linking, you have a pointer to an item in a list that contains pointers to another topic. In soft linking, there are no pointers. Rather, there is the information required for semantic discovery based on subject affinities.

Soft linking and list generation

One of the things you may have realized about the soft-linking approach is that when you run a query on a particular subject expressed by subject-affinity markup, you may get more than one result. Since conventional Web links only point to one page, what do you do when you get multiple results?

For an answer, we must look back to the discussion of lists in Chapter 5. As we noted there, lists are a common feature of the Web. On the Web there are almost always multiple resources on any subject, so when you look for topics based on a subject affinity, you will probably get back a list of resources. The question is, how do you present that list.

This situation occurs frequently on dynamically generated websites. Amazon's pages are full of generated lists, and the page layout is designed to accommodate them. So one way to deal with multiple resources is to create a list of links on the page. In the middle of a passage of text, you might prefer to have the list presented as a pop-up of some kind, triggered by a link on the original phrase in the text.

A third option is to create separate list pages. List pages can serve other purposes, too. For instance, people may land on them as the result of a search or may wish to bookmark a list page on a frequently visited subject.

CHAPTER 21

Reuse

Reuse is the hot topic in content strategy and technical communications today. So how does writing and managing Every Page is Page One topics fit with reuse?

We should begin by recognizing that reuse means different things in the paper world than in the Web world. In the paper world, reuse generally refers to the practice of taking one piece of source content and using it in many different publications. Thus the same content may be used in a user's guide, a quick reference card, a marketing brochure, a press release, and a product package. Additionally, content may be reused in the manuals for multiple related products or successive versions of a product.

This approach can bring sizable benefits in terms of reduced writing time, greater consistency, ease in finding and fixing errors, and savings in translation. It can also come with significant costs for content management, training, and overhead in the writing process. To offset these costs, a common goal is to maximize the amount of reuse in the system, thus increasing the cost savings. To accomplish this, companies will often break down content into very small pieces, the building-block topics we talked about in Chapter 6.

These building blocks are often smaller than Every Page is Page One topics, so if you want to make Every Page is Page One topics from building blocks, you need to plan in advance. Failure to plan deliberately can lead to the fragmented outputs – the gears and screws rather than alarm clocks – we talked about in Chapter 19.

Reuse on the Web

Placing the same content in many publications can present a problem in the Web world. One of the key things to remember about the Web, or even your own help system, is that it is a flat information space. Duplication makes sense in the paper world, because each paper document lives in its own little valley with steep mountains between it and the next document. We duplicate content between paper books so readers don't have to climb the mountain between one book and the next.

The Web is not a valley. It is one vast flat plain. By default, any search searches the whole thing. Using paper-style reuse on the Web puts multiple identical, or near-identical, items into a flat information space. Search engines don't like that.

As Peter J. Meyers comments:

> One of those [SEO] issues is duplicate content. While duplicate content as an SEO problem has been around for years, the way Google handles it has evolved dramatically and seems to only get more complicated with every update.
>
> —Peter J. Meyers, "Duplicate Content in a Post-Panda World"[1]

I won't go into the technical details – read Meyers' blog if you want to know – but the bottom line is, duplicate and near-duplicate content is bad for SEO (Search Engine Optimization). At least some of your duplicate content won't show up in search, and duplication may prevent search engines from thoroughly indexing your site.[2]

More importantly, even if each piece of duplicate or near-duplicate content is indexed by Google or your internal search engine, the chances of a search returning a link to the content that is right for the reader's purpose or version of the product is inversely proportional to the number of variants you have. Having a navigable context can help a little here, but the better solution is to avoid the duplication.

Jakob Nielsen sees content reuse on the Web as something users, rather than authors, do. In his post "Write for Reuse"[3] he says, "People will use your copy differently than you expect, and you should try to write with this fact of online life in mind." His principle guidelines for writing content that users can reuse for different tasks are, "assume your information will be used out of context," "modularize your information," and "use specific language," all things that a good EPPO topic should do.

I've said that it is best to think of the Web as a giant multifaceted information filter that dynamically groups content by subject. Therefore, the Web reuses content all

---

[1] http://moz.com/blog/duplicate-content-in-a-post-panda-world

[2] What is good or bad SEO is both volatile and subject to much debate. Things may be different by the time you read this. But getting a set of near-identical search results is never particularly helpful.

[3] http://www.nngroup.com/articles/write-for-reuse/

the time, as search and social curation groups content into semantic clusters according to the needs and interests of different readers.

Individual organizations can enhance the natural filtering action of the Web by making their content available through an API. A content API makes content available as structured data objects that people can request in different forms and manipulate in any way they want. Publishing through an API makes your content available for reuse in any number of ways. Sara Wachter-Boettcher provides good coverage of this in her book *Content Everywhere* where she comments:

> Like the oft-touted NPR example, you may also need to get content out across lots of very different products or destinations. Or, you might not really be dealing with publishing content all at once, but rather creating a repository of information that can be used in the future, however it's needed—or even selected and arranged by your users, creating their own personalized collections of your content to suit whatever their needs happen to be. This is the world of reusable and reconfigurable content: content that can be pushed out to lots of places at once, assembled and associated with other relevant bits on the fly, displayed in different combinations for different purposes, or connected and combined by users themselves.
>
> —*Content Everywhere*[26, pp. 2505–2512]

The big difference here is that whereas the emphasis in the paper world is on authors reusing content themselves (usually by hand) to create multiple static publications, on the Web, the emphasis is on enabling readers to reuse content dynamically for themselves using automated means.

In some cases, these Web-based systems require a log in – they need information about the user in order to personalize the content. This means content is not directly visible to search engines, which avoids the SEO problems raised by exposing multiple similar pages. However, such a system relies on users being willing to log in – to abandon the richness and scope of the Web for the sake of the customized view a single site can provide. That may be a tall order, and you need a well-thought-out content strategy to be confident of success using this approach.

One of the implications of this emphasis on automated reuse on the Web is that any unit of content offered for reuse either must be a whole unit of content or must have enough metadata attached to it that it can be built into or included in a whole unit of content based on the metadata alone, without human intervention.

Static vs. dynamic reuse

There are, essentially, two forms of reuse, which we might label static and dynamic.[4] With static reuse, you create a piece of content once and then assign it one or more distinct roles. Later you, or someone else, can assign it to a different role, perhaps assigning it several distinct roles [5] in different publications on paper, electronic media, and the Web.

Static reuse is generally done by hand. In the case of DITA, the system most commonly used for reuse, it is done by creating DITA maps.[6] You can use a DITA map to build an EPPO topic, a book, a help system, or a Frankenbook.

Dynamic reuse, on the other hand, is more like creating a single piece of content to play a single role that allows it to appear in multiple places. There is no separate human analysis or decision involved when a topic turns up in each of the places it occurs. It occurs because its single and original role demands it.

Building-block topics created for static reuse may not be suitable for dynamic reuse. We have looked at a few examples of topics in help systems that clearly were not self-contained in the EPPO sense (Figure 4.3, Figure 10.1). Because of the table of contents in the left-hand pane of most help system, there is some hope that readers can make sense of those building-block topics. But imagine if one of these topics were called

[4] Ann Rockley and Charles Cooper use the terms "manual" and "automated" reuse in a very similar sense. [24] I prefer the term "dynamic" because it encompasses the reuse that happens as a result of the dynamic semantic clustering action of the Web, without any action or intention from the originating organization.

[5] By *role*, I mean that a topic might start out as the product overview in a brochure and subsequently be assigned the role of product overview in a getting-started guide. Of course, changing roles doesn't mean changing topic types. In both roles, this topic would be a product overview.

[6] This is not to say that you can't generate DITA maps automatically from metadata. Some DITA implementations do exactly that.

from a content API and presented apart from its TOC. It would probably be of little use to a reader.

An Every Page is Page One topic, then, is a more natural unit to offer through a content API. Because EPPO topics are self-contained and establish their context, they should still make sense and be useful no matter where they are displayed. (In this sense, it does not matter if EPPO topics end up being found and grouped with other topics by search, social curation, or an API.)

This is not to say that more fine-grained data can't also be delivered through a content API, as long as it can be used to build an Every Page is Page One topic for the reader. If your EPPO topics are created using subject-specific computably structured writing, you can manipulate them behind the API or allow users to manipulate content themselves. You can use this kind of manipulation to pull in additional information or adapt the content to different contexts.

Even if you are simply delivering Web pages and not a content API, you may still want to take a different approach to reuse than you would in the book world. On the Web, linking to a single instance of content usually makes more sense than duplicating it in multiple places, and it makes search engines much happier.

This does not necessarily mean that you have to choose one style of reuse over the other. With the right structured-writing approach it may be possible to do paper-style reuse for paper outputs and Web-style reuse for Web outputs from the same source. You may have to make some choices about which is your primary medium – it is hard to make the same content source create EPPO topics for the Web and a consecutive narrative for paper – but you should still be able to use a reuse strategy appropriate to the target media, rather than using the same strategy for both.

Other forms of reuse

There are other forms of reuse besides deploying topics to more than one publication. Variable substitution, similar to that used in a mail-merge application, can be used to produce different variations of a topic. Conditional text markers can be used to hide or reveal different pieces of text within a topic based on certain conditions.

On paper, these approaches involve creating multiple similar topics to deploy across multiple publications. On the Web, it can be done the same way, or you can resolve the variables and conditions on the fly, using information you already have about the viewer or choices made by the viewer. Given the SEO problems that duplicate content can create, the latter is probably preferable. These practices work the same with EPPO content or any other type of content.

Reuse, linking, and interactive pages

As noted above, linking can be an alternative to reuse. For instance, in the book paradigm you might include task information in multiple workflow descriptions. In EPPO, you could handle the same situation by writing multiple workflow topics, each of which linked to the specific task or procedure topics.

However, the distinction between linking and reuse can get blurred. If the content that users see is simply a static file served from web server, the distinction between reuse and linking is clear. Either the content is included in multiple files or it is included in just one and other files link to it.

But with interactive content, which can include both pages with interactive features created in JavaScript and pages generated interactively by a content API, content can be pulled into the page, either when the page is loaded or when a link is clicked. In this case, if you are reading a workflow topic and you click on the link to a specific task or procedure, the link might result in the task or procedure being shown inline rather than by moving you to a new page.

These techniques can be used in the context of any information design, including Every Page is Page One.

CHAPTER 22
Making the Case for Every Page is Page One

Throughout this book, I have been making the case that Every Page is Page One is the right information design pattern for most technical communication applications. The heart of that argument is that people learn through experience and through information acquired in the context of experience, and that they want short pieces of content that serve their specific purpose.

This has always been the case, and the systematic textbook model has never been the right one for technical communication. But the advent of the Web has made the need for an Every Page is Page One information design more acute by reducing the distance between information sources to zero and allowing information foragers to move effortlessly from one information source to another. Readers now consume our content in the context of the Web, with habits and expectations formed by the Web. The textbook model, never optimal, is now nonviable.

Readers have moved away from the traditional products of technical communication to the social technical communication environment of the Web. The move is not entirely uniform – audiences in some product sectors have moved faster than others, but as computing moves off the desktop and becomes ever more ubiquitous, the move will happen in every product sector. Changing to an EPPO information design and putting your technical communication on the Web are ways to get back the audience that technical communicators are losing.

ROI calculations are often stated as if the present were a stable baseline and you could project future gains against the cost of change. But technical communication today does not sit on a stable base. The ROI of any change you make in a technical publications organization today is first and foremost about not losing ground.

The ROI of any change you make in a tech comm organization today is first and foremost about not losing ground.

That is the macro case for Every Page is Page One. But there remain a large number of practical concerns to be addressed. This chapter will look at how to make the case for Every Page is Page One in terms of the practical concerns and everyday challenges of a technical publications group.

EPPO and resource constraints

Over the last several years technical communications departments have been increasingly challenged by tightening resource constraints. Budgets are tight and tech comm head count growth is seldom allowed to keep pace with engineering head count growth. Technical communicators need to find ways to work more efficiently.

Producing nearly any complex product in small units is more efficient than producing it in large units, largely because small units are easier to create, errors are found sooner, and production can flow more evenly.[1] Working in smaller units also allows for better resource allocation. In particular, it allows writers to specialize in certain technology areas, enabling them to produce better content more quickly across several products instead of having to write everything about one product.

Reuse may provide further savings. As noted in Chapter 21, creating EPPO topics will open up possibilities for reuse between different product lines and different generations of product. Using features available in your current tools, you may be able to use variables and conditional text in your EPPO topics to allow you to use those topics more widely. While you might not achieve the same percentage of overall reuse that you would with a full CCMS and building-block topics, you may be able to achieve significant savings without the major investment that a more fine-grained reuse strategy would require.

Shortened product cycles are another form of resource constraint. When products are developed more quickly and released more often, less time is available to develop content. Again, working in smaller units, particularly units that are loosely coupled and don't require elaborate assembly, can help you manage shortened product cycles.

---

[1] For more on this, see *Lean Thinking*[29] by James P. Womack and Daniel T. Jones.

By allowing you to work in smaller units, EPPO allows you to turn content around more quickly. A single topic can be researched, written, reviewed, approved, and delivered quickly. Working in topics allows you to avoid the log jams typical of trying to bring a book to a publishable state. Rather than a massive review period at the end of the product cycle – just when the people you are asking to do the review are busiest with their own work – you can spread the reviews out over the whole development period. This leads to a more evenly paced content development process and, consequently, less stress and higher quality.

All of these benefits can be realized without major infrastructure investment simply by designing and building content in smaller units.

EPPO and continuous delivery

Much of the useful information about how to use, connect, troubleshoot, and fix a product is developed after the product is released. This has become even more true in the age of the Web. Where previously most products stood alone, today, everything from software to consumer electronics to PCs to industrial robots is connected to databases and to other software and machinery over intranets and the Web.

More than ever, the issues that people struggle with are not, "how do I make this program or device work by itself," but "how do I get it to talk to everything else." And since many of the things that users are trying to connect to were released after their own device or program, there is no possible way to answer those kinds of questions in a document written and finalized before the product was even released.

Traditionally, technical publications have provided only information that was available before the product was released. This was primarily because of the practical constraints of having to deliver documentation physically with a product in the form of paper manuals or computer media such as a CD-ROM. After the product was released, except in cases where updates were required by regulation or contract, delivery of newly discovered information was handed off to the support department. (Interestingly, the knowledge bases created by support are often similar to Every Page is Page One format, though they often lack formal structure or links.)

Multiple factors make this deliver-once-at-release-time model nonviable:

- Customer expectations are now driven by the Web. Customers expect documentation to always be up to date with the latest known information, and they see no reason why they should have to look in two unconnected sources (the documentation and the knowledge base) for information about the same product.
- Software products (and hardware products such as phones and tablets that are largely animated by software) are increasingly moving away from the major release model towards a more continuous roll out of features. Information needs to be updated on the same schedule.
- The growth of software as a service means companies can roll out new services as soon as they are ready. Again, information delivery has to keep pace.

For any of these types of continuous delivery, Every Page is Page One topics are ideal. Because Every Page is Page One topics have no sequential dependencies, they are essentially plug and play. You can add and remove topics from the content set at any time without disrupting anything. (You will, of course, need to manage link dependencies when you do this.) Every Page is Page One handles continuous delivery better than building-block approaches, which require you to assemble and sequence building blocks into larger and/or hierarchical information products for each delivery.

EPPO and content change

Keeping up with changes is an ongoing problem for technical communicators. This is not just the problem of updating content from release to release. Most product documentation was originally designed for version 1.0 of the product, when it was fairly small and simple. But products grow over time. New features are added, and the product is adapted to fit new uses. The structure that made sense for documenting a small sapling is no longer right for documenting a mature tree. But season after season of growth and deadlines leaves few opportunities for a major overhaul.

I noted in Chapter 12 that books produced by tech pubs departments tend to change levels more than third-party books. This is mostly because they are maintained over multiple generations by multiple authors who must accommodate the demands and suggestions of countless developers, product managers, and field staff, all of whom insist that some piece of information that has just come to their attention absolutely must be included in the documentation.

Technical writers rarely have the time and opportunity to design a book with a mature holistic view of the subject matter. Even if you own the whole book and start from scratch, as opposed to updating an existing book, you are always documenting a moving target against an unreasonable deadline. This is not a work environment that is conducive to creating a brilliantly designed curriculum.

Of course, more often you inherit books that have become disorganized over time or whose original organization is no longer appropriate for the size and scope of the current product. Everybody involved, from writer to manager to project manager, recognizes that a reorganization is desperately needed, but there is never time in the product cycle to do a complete reorganization, and starting and not finishing can leave the situation worse than it was to begin with. So the problem – including the multiple, abrupt, and inconsistent changes of level – continues to get worse and the manual becomes less and less usable with every iteration.

On the other hand, with Every Page is Page One topics you are always working with a manageable unit. There may not be time in the cycle to fix all the topics, but there is always time to fix the topic you are working on. You don't have to wait for that big chunk of time that never comes to reorganize the whole. You can keep individual topics in good shape, and if you get a spare day or a spare week, you can make progress on the backlog of topics that need fixing. Working topics will allow you to make steady progress on improving the quality and organization of your content.

There may not be time in the cycle to fix all of the topics, but there is always time to fix the topic you are working on.

EPPO and content aging

One of the biggest content management problems for documentation sets and websites alike is content aging. At a certain point, content becomes outdated and redundant and should be removed. If it is not removed, it can confuse navigation, compromise the findability of current content, and mislead the reader. However, removing obsolete content is not simple. In sequentially or hierarchically organized content, obsolete content is often tangled up with current content. Simply removing the obsolete content can leave holes in the organization or broken links between pages. Detecting and repairing these holes can significantly increase the cost of removing obsolete content, which often means that the task gets put off, resulting in obsolete content becoming even more entangled with good content and making the problem of removing it even more daunting.

Every Page is Page One information design helps alleviate this problem by reducing or eliminating the tangles between old and new content. Of course, an Every Page is Page One topic does link richly to other topics and other topics link to it, and these tangles have to be dealt with. But if your linking strategy is based on subject affinities and soft links, the links between topics do not create hard dependencies. You can remove a topic from the set and create new links in the topics that link to it, basing those new links on the subjects and subject affinities of current topics. If you use the soft-linking technique described in Chapter 20, this will happen automatically, and you can simply remove the obsolete topic without worrying about the tangles.

For example, consider a piece of content that contains a link to your annual developer's conference. Ideally, that link should point to the page describing the next developer's conference. If that content is created with a hard link, it will be obsolete in a year's time. Instead of using a hard link, it would be better to record a subject affinity, such as the following:

```
<event>annual developer's conference</event>
```

Your build system can then automatically generate a link to the current developer's conference page.

To ensure that the correct developer's conference topic is selected automatically, you can add metadata to your event description topics to specify the name of the event and the date. The build system will then have a rule that says: always link instances of events mentioned in the content to the event topic with the nearest date in the future (or, if there is no date in the future, to the nearest date in the past.)

With these rules in place, simply adding an event description topic for next year's developer's conference will cause all the marked-up mentions of that event to link to that page (unless this year's conference has not taken place yet, in which case they will keep pointing to this year's conference until it is over and then immediately start pointing to next year's conference.)

Another aspect of content aging is identifying topics that are obsolete. Fundamentally this depends on the metadata attached to a topic when it is authored. Metadata should include the information needed to implement an appropriate content aging strategy. Every Page is Page One does not make any special contribution to defining and managing such metadata. However, Every Page is Page One does help ensure that topics fully merit the metadata that is applied to them (Chapter 19), which means that you can rely on the metadata with greater confidence when implementing an aging strategy. This could have significant benefits. For instance, in concert with a soft-linking strategy, it could provide you with enough reliability to automate content removal completely, so that content is simply removed as soon as its expiration conditions are met, without human intervention.

In our developer's conference page example, the old developer's conference page could be removed by an algorithm as soon as two conditions are met:

- The event described is in the past.
- A page describing the same event, but with a date in the future, exists.

EPPO and agile methodologies

The adoption of agile methodologies by software development organizations has been challenging for many technical communication organizations. Often the timing and needs of technical communicators are not taken into account when organizations adopt an agile methodology.

The usual model for incorporating technical communication into the agile process is to assign a writer to the scrum team and then define documentation requirements for each sprint. But while a switch to agile may involve significant changes in the way software is designed and delivered, there is usually no such provision made for changing the way documentation is designed and delivered. Technical writers are often left with their old deliverables, their old tools, and a schedule and methodology dictated by the development organization.

Organizations that use agile processes should be making frequent deliveries to customers to get the feedback necessary to guide product development. And documentation should be delivering to customers at the same time to assist customers during the trial and to get feedback on the documentation. However, for documentation groups still using tools and processes designed to support a one-time, final-release publishing schedule, agile development only exacerbates the problems most groups are already encountering with continuous delivery processes.

Technical communication groups must change their processes and deliverables to thrive in an agile environment. Instead of trying to adopt the development group's agile process, I suggest that technical communication groups should look into developing their own lean[2] development process and integrating it with the development group's agile process. However, even for groups that choose to work within development's agile process, moving to EPPO topics will make it easier to operate effectively in an agile environment.

In an agile process, you can deliver EPPO topics to each sprint without having to worry about how they will fit into a larger structure. And if additional documentation requirements emerge after release, you can add EPPO topics to the documentation set without the need to restructure or reissue it.

One of the features of Agile development is that you do not spell out every detail of the design before development begins. This is based on the realization that trying to define all the features of a system up front usually leads to over-specification of some requirements and omission of others. Developing a system in multiple iterations with frequent delivery to customers is a strategy for avoiding the problems associated with traditional, up-front specifications.

Instead, the development process is designed to flush out user needs by giving both the development team and customers experience with working prototypes. The par-

---

[2] *Lean* (http://www.lean.org/whatslean/) is a set of principles aimed at eliminating waste in production processes. Though of independent origin, agile could be framed as the application of lean principles to software development, and there is increasing cross-pollination of ideas between lean and agile methods (http://en.wikipedia.org/wiki/Lean_software_development). Whereas agile is specifically designed for software development (and is not necessarily optimized for content development), lean is more general and can be applied to many different processes, including content development.

allel with John Carroll's findings on the paradox of sense making (Chapter 14) should be noted here. Users of a new system require direct experience to adjust their mental mode. People who specify a new system have a similar need for experience.

This same process of deliberate discovery of requirements through the development process applies to technical communication as well (and is why you should deliver to customers on a regular basis, just as development does). This means you will be working without a complete, detailed documentation plan (there's no complete, detailed product plan to base it on anyway).

Because Every Page is Page One topics are, ideally, organized bottom up, managing subject affinities is an important part of developing EPPO content. Once you start managing subject affinities, you will quickly discover how powerful an aid they are to planning content requirements and to maintaining and developing a content plan through an agile development process.

Traditionally, we tried to plan our documentation sets from the top down, compiling lists of the topics to be written or outlining tables of contents, but even the best effort from the top down will miss things. As soon as you start managing subject affinities, you will start finding holes in your content. As we author topics we naturally refer to other subjects. However, we often don't realize that we have no documentation on the subject we just referred to. Managing subject affinities lets you collect and review all the subjects your content mentions and compare that list with the list of the subjects you document. This quickly shows where things are missing.

Careful management of subject affinities will help you in your planning process and help you find omissions faster.

EPPO and content management

You can start creating Every Page is Page One content with your current tools today, without spending a cent on new systems or interrupting your current workflow. You can do it using the word processor, desktop publishing system, help authoring tool, wiki, or Web CMS you use today.

At the same time, every tool has an inherent bias towards a particular kind of information design. People don't buy tools that are perfectly general, because a perfectly general tool is not optimized for the work they are currently engaged in. Tools that succeed in the market are the ones that are optimal for current designs and processes. Therefore, all tools are optimized for a particular design and a particular process. Nowhere is this more evident than in the content management space, where we find content management systems of every kind and description, all managing content and each optimized to manage a particular type of content and a particular content development process.

This does not mean that you can't use such systems to produce a different design using a different process. But tools will not work optimally when you try to use them for tasks they weren't designed for. A CMS that is designed for managing building-block topics and for assembling them into books and help systems needs to support the collaboration and discovery required for this approach. By contrast, Every Page is Page One topics have far fewer dependencies than building block topics and, therefore, don't require these features, but they do benefit greatly from features that support automated linking and bottom-up organization.

Tool change is in the air in the technical communication world. There are more and more tools, and more and more types of tools, vying for our attention, and more and more voices are joining the chorus to say that technical communication departments have to move to a content management system.

Switching to a tool that better supports your new information design, either after you change design or at the same time, makes a lot of sense. The danger is that a department may succumb to the pressure to change tools, especially to the drumbeat of CMS advocates, without fully considering what kinds of information designs and delivery schedules they will need in the future. For instance, if you switch from a desktop publishing tool to a CMS while continuing to produce manuals and hierarchical help systems, you will likely choose a CMS with a bias towards those kinds of outputs. If you subsequently switch to an EPPO information design, you may find the CMS is not optimal for that purpose.

Your delivery process is also a vital consideration. Just as tools have a built-in bias towards certain types of outputs, they also have a built-in bias towards certain types of delivery schedules. A typical desktop publishing system is built on the conventional model that publication is the culmination of a long development and approval system. A wiki, on the other hand, is built on the model that content is always live and can be edited by anyone at any time. A desktop publishing system may be able to support more frequent publishing and a wiki may be able to delay and review content before making it public, but process bias is built into the architecture of each.

Therefore, if you are planning to move to EPPO information design or are planning to move to Web delivery, don't buy new tools before you have a new information design and delivery method firmly in mind.

EPPO and PDF/help

Many technical communication organizations still need to produce PDF and help output. As Alan Pringle commented in a recent blog post, you should not abandon PDFs just because they are no longer cutting edge.

> Even if you don't have legal reasons to continue to provide PDF files, it's the height of hubris (and stupidity) to assume your customers will immediately accept content distributed in new ways. Instead, be smart by offering your customers choices in how they consume content.
>
> —"'No PDF for you!' The destructive power of arrogant thinking"[3]

At the same time, it is worth investigating seriously whether the demand for PDF is real or not, as JoAnn Hackos relates:

> We know that people will tell you that they prefer PDF. One of our speakers at the Best Practices conference two years ago, Bob Lee from Symantec said they had done a study of those customers who said they preferred PDF, when they offered them both PDF and HTML-based topic-based assistance online

[3] http://www.scriptorium.com/2013/06/no-pdf-for-you-the-destructive-power-of-arrogant-thinking/

> they found that even the people who said they preferred PDF all actually selected the topics and the HTML output at a ratio of 26 to 1. So lots of times when you get that feedback that people prefer PDF, it is mostly because it is the only thing that people know to tell you.
>
> —"DITA Rockstar Summer Camp – DITA and Minimalism"[4]

Hackos observation aligns with John Carroll's finding that users asked for information in one form, but actually used it in quite a different way. We are not always great analysts of our own preferences and behavior patterns, and sometimes we ask for the conventional thing with the familiar name rather than the thing that actually works best for us. As Steve Jobs famously said, "It's really hard to design products by focus groups. A lot of times, people don't know what they want until you show it to them."[5]

One of my clients investigated why customers were still asking for PDFs and discovered it was because the alternative presentation they were offered, which was an Eclipse help system, was too hard to search, and also because many users did not use the company's GUI development environment and, therefore, did not have access to the help system. PDF was not ideal for them, it was just preferable to the only alternative they were offered.

Another factor, I believe, is that people prefer to consume content in the format it was designed for. If you are writing your content as books and your help or Web content is being created by mechanically separating those books into topics and compiling them into Frankenbooks, chances are people will find a PDF easier to use because it keeps the linear structure of the content intact.

So, it may well be that your users are demanding PDFs today not because they inherently love the PDF format but because it is all they know to ask for, because it is better than the alternatives you are offering, or because it is a closer match with your current information design. If you moved your content to an Every Page is Page One information design and offered it in a medium with good support for search and bottom-up navigation, they might very well stop using PDF and they might eventually stop asking

---

[4] http://www.youtube.com/watch?feature=player_embedded&v=3wSybrSeEYQ

[5] https://www.helpscout.net/blog/why-steve-jobs-never-listened-to-his-customers/

for PDF. However, until you are reasonably certain that that time has come, it may be wise, as Pringle suggests, to keep producing PDF.

The question then becomes, are customers asking for a PDF because they want a linear manual? That's a harder question to answer for certain, beyond the answer I gave in Part I, "Content in the Context of the Web," which is that readers living and working in the context of the Web generally prefer Every Page is Page One content. There are definitely some people who continue to prefer forms of content organization from the book era, so it is legitimate to ask if such people make up an inordinately large proportion of your product's demographics. But even if you answer in the affirmative, how much longer will that continue to be the case?

Most of us are now thoroughly habituated to the Web and to the way the Web organizes information. Certainly the upcoming generation is unlikely to take any comfort in the structure of a book. As Brian S Hall laments:

> I'm not sure who to blame. His mother, perhaps, or the public school system. But it turns out that my son—days away from graduating from high school—does not know how to send mail through the U.S. Postal Service.
>
> I am not making this up.
>
> The boy has a smartphone, a tablet and a laptop, does some basic coding, is pretty good at computer-assisted design and gets excellent grades. He can bang out what appears to be 60 words per minute using only his thumbs. But a letter? Forget about it—he doesn't even know how to properly address an envelope.
>
> —"My Teenage Son Does Not Know How To Mail A Letter, And I Blame Technology"[6]

If paper plays such a small role in this boy's life that he has not learned how to mail a letter, he probably does not expect the documentation for his laptop to be structured like a book.

[6] http://readwrite.com/2013/05/27/my-teenage-son-does-not-know-how-to-mail-a-letter

In short, whether people still want PDFs and help, and whether our companies still want to keep their documentation off the Web, there is no longer any reason to think that our readers are culturally wedded to the old systematic textbook format of a user manual. Even the studies John Carroll did in the 1980s showed that people did not use manuals as they were designed to be used.

If we are no longer culturally tied to the textbook form and if we know it has not worked for at least the last quarter century, we don't need to be afraid to walk away from it.

So, if we are no longer culturally tied to the textbook form and if we know it has not worked for at least the last quarter century, we don't need to be afraid to walk away from it.

If you need to, you certainly can build PDFs and classic tri-pane help systems from Every Page is Page One topics. The details will depend on the tools you use, but as an information design pattern, Every Page is Page One works as content for both help systems and PDF manuals.

There is actually nothing at all new about building manuals out of Every Page is Page One topics. There is also nothing uncommon about building books that are not designed to be read sequentially. I mentioned one such case in Chapter 4, the *Popular Mechanics Complete Car Care Manual*, which is a collection of car-care articles, all of which presumably appeared in the magazine. It has a table of contents that is organized by the parts of the car to help readers pick the articles that describe the tasks they want to perform.

That is essentially all there is to making a manual out of Every Page is Page One topics. Choose a set of topics to include, figure out which aspects of their subjects make the most sense as static groupings, and create a table of contents. Programs like Word or FrameMaker make it easy to do this. You simply use the book-building feature to generate a TOC and index, add some front matter, and you have yourself a book.

Avoid the temptation to add linking text or string the EPPO topics into a narrative. It isn't necessary. Few, if any, will read the book through (even those who say they will, as Carroll found). Even if they do, most people skip the connective passages anyway. You don't need to make a narrative. People don't read manuals that way.[7]

---

[7] They do read other sorts of books that way, of course, so if you genuinely need a single long sequential narrative, that is what you should write. I'll say more on this in Chapter 23.

If you have big picture topics and pathfinder topics, you might want to group these at the beginning of the TOC. This is not because you expect people to read them in that order, but because these types of topics can also be useful in helping people orient themselves so that they can dive into the rest of the content in random order.

Just as they can be assembled into a book, EPPO topics can also be assembled into a hierarchical help system. Of course, those topics won't depend on the TOC in the same way that hierarchical topics do, but that is a good thing, since readers generally won't read them that way. You should, of course, maintain the subject affinity links between topics in the help system to facilitate bottom-up navigation.

EPPO and content marketing

From a technical communications point of view, putting technical content on your website is clearly a good content strategy because the Web is where people look for answers to technical questions. But unlike paper, the Web is a flat information space, and you cannot easily hide information from one audience while revealing it to another. What you show, you show to everyone. Different people use different filters for different purposes, so people with different agendas see different content. However, when you put your documentation on the Web, it becomes an element of your content marketing strategy, for good or ill.

Unlike paper, the Web is a flat information space, and you cannot easily hide information from one audience while revealing it to another.

One immediate benefit of putting technical content on the Web is that you will have more pages for search engines and information seekers to find. But is that always as good thing? If content is what attracts people to your site, then presumably more content should be a good thing. And if we bear in mind what we learned about the long tail phenomenon in Chapter 2, it would seem that having a lot of content should help attract more customers because they can meet more of their content needs in one place.

Yet many content strategists take the opposite view, urging people to radically reduce the amount of content on their websites. Gerry McGovern writes:

> Driven by an ill-advised content focus, marketers and communicators are producing huge quantities of fresh content news and articles. This may indeed deliver short-term traffic boosts. But today's news is tomorrow's clutter. Websites are becoming harder and harder to navigate and search because there is so much stuff on them....
>
> Invariably, when we delete 90% of an organization's website, sales go up dramatically, customer support requests go down and overall customer satisfaction goes up.
>
> —"Communications and marketing professionals at a crossroads"[8]

And yet, I imagine that if you told Jeff Bezos that Amazon would do better by removing 90% of its pages; he would laugh in your face. Similarly, if you told Jimmy Wales that Wikipedia would be a more popular and useful resource if 90% of the articles were removed, I'm sure he would disagree. Thus we have a conundrum.

Part of this conundrum can be explained simply by observing that a lot of stuff on company websites is junk. The Web is a filter, so some of the junk will get filtered out. But it is not a perfect filter. Some of the junk still gets through. Worse, having a lot of junk on your site can cause the good content there to get filtered out. This could explain McGovern's observation that sales go up when excess content is removed. In this case the junk is actually getting in the way. It clogs the filters:

> Customers are finding it harder and harder to buy and complete tasks because they end up on these 'getting attention' content pages, whereas they were trying to find the homepage for product X.

In other words, the junk is crowding out the good stuff – the pages on which people can actually act and buy. But it cannot be mere quantity of content that causes this crowding or long ago Amazon would have reached a limit of growth where adding more pages would make it harder for people to buy. That has not happened. In fact, quite the opposite has happened. Amazon continues to serve the long tail with even more obscure titles, and the demand for those titles has only increased.

---

[8] http://gerrymcgovern.com/new-thinking/communications-and-marketing-professionals-crossroads

Of course, every page on Amazon is a place to buy. You are almost never on a page on Amazon where you can't click to buy something. But that's not the point. There are millions of pages on Amazon for things that you don't want to buy and only a few for things you do want to buy, yet people find the pages for the things they want to buy and they buy them. Similarly, for Wikipedia, adding more articles doesn't seem to make it harder to find the article you want. If anything, it may make it easier as there are more links available to help you get there.

A clue to understanding this conundrum may come from Bruce Tognazzini's recent post about the Apple UI, "The Third User, or Exactly Why Apple Keeps Doing Foolish Things,"[9] where he observes:

> Apple keeps doing things in the Mac OS that leave the user-experience (UX) community scratching its collective head, things like hiding the scroll bars and placing invisible controls inside the content region of windows on computers.
>
> Apple's mobile devices are even worse: It can take users upwards of five seconds to accurately drop the text pointer where they need it, but Apple refuses to add the arrow keys that have belonged on the keyboard from day-one.
>
> Apple's strategy is exactly right—up to a point. Apple's decisions may look foolish to those schooled in UX, but balance that against the fact that Apple consistently makes more money than the next several leaders in the industry combined. While it's true Apple is missing something—arrow keys—we in the UX community are missing something, too: Apple's razor-sharp focus on a user many of us often fail to even consider: The potential user, the buyer.
>
> —Bruce Tognazzini

In focusing on the buyer, Tognazzini explains, Apple is not focusing on making its products as easy to use as possible, but on making them *look* as easy to use as possible to the person looking at them in the store. To do this, they hide all kinds of useful user interface elements to make the product look easier to use, even though hiding them actually makes the product harder to use.

[9] http://asktog.com/atc/the-third-user/

It remains to be seen whether this approach will be sustainable as phone and tablet buyers become more sophisticated and more focused on functionality, but the observation goes a long way towards explaining the conundrum we are looking at. A site with a lot of rich content may indeed be the most useful and may contain the most value for the widest array of customers, but if it looks big and complicated and if it is hard to navigate, it can scare off buyers.

Given this, is putting technical content on the Web going to make your product look more or less attractive? Two things are pretty clear.

- You don't want the technical content to get in the way of visitors accomplishing their tasks, including buying stuff.
- You don't want to make your company's product look hard to use, which is exactly what you are doing if you show them something like Figure 22.1:

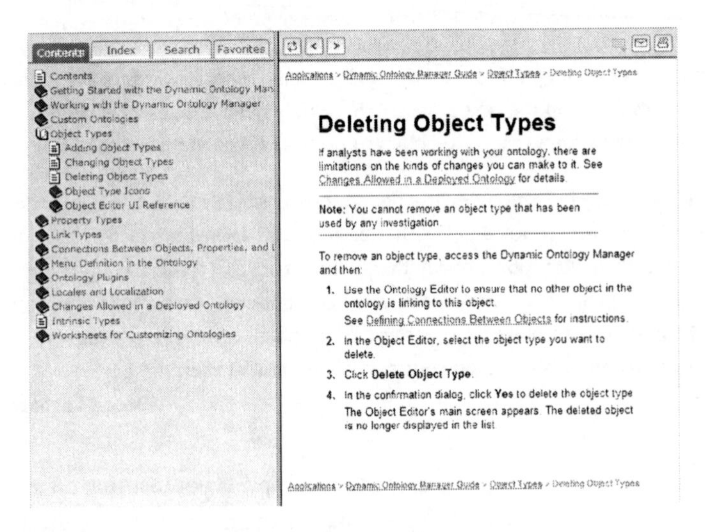

Figure 22.1 – Big docs make products look hard to use

Putting up a traditional help system, with a table of contents on every page, rubs the user's nose in just how big the documentation set is. While there are a few people who revel in this kind of thing, by and large it is a big turn off.

One of Apple's early Mac commercials hinged on exactly this point. It started out with a picture of a PC and the narration went like this: "This is a highly sophisticated office computer, [shows the PC] and to use it all you have to do is learn this. [drops three heavy binders on the table beside it] This is Macintosh from Apple, [pans to the Mac] also a highly sophisticated office computer. To use it, all you have to do is learn this. [drops a few sheets of paper]"[10]

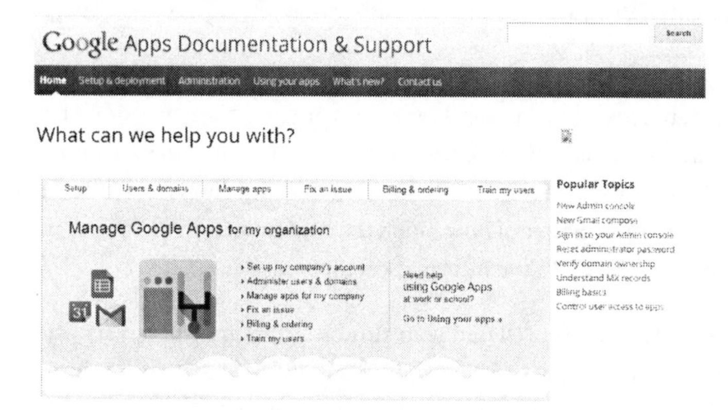

Figure 22.2 – A less terrifying documentation home page

But there is a conundrum here as well. People want answers to their questions, and they will complain if they can't find them. Seeing the doc set as a whole scares them. But they want little bits of it now and then, and each person wants a different little bit. So you need a documentation set that seems small and simple (to avoid scaring off buyers) but is actually comprehensive (so answers are available when needed).

How do you do that? You do it the way Amazon and Wikipedia do it – with bottom-up navigation. Bottom-up navigation never overwhelms readers with how big the information set is, it just makes it easy to navigate along any line of subject affinity.

Contrast Figure 22.1 with Figure 22.2. Does Figure 22.2 give you any sense that there is a huge documentation set behind it? No, it makes everything look safe and simple.

[10] http://www.youtube.com/watch?v=VaZgtQRmunA. You can find the commercial at the 2:35 mark in this compilation on YouTube.

This documentation set is actually pretty big, but once you're inside, you will find plenty of links to get you from where you are to where you need to be.

An Every Page is Page One information design supports bottom-up navigation that can hide how big the documentation set is. No matter where readers land, there are ample links to take them to related pages. Yet readers are never hit over the head with the full extent of the documentation set or (worse yet) expected to navigate it.

Clutter is not a product of the amount of content on your site.

Every Page is Page One topics are designed to meet specific needs. Perhaps more importantly, because they support bottom-up organization, they don't have to create website clutter that makes it hard for users to do what they want. Clutter is not a product of the amount of content on your site. Wikipedia is not cluttered nor is Amazon. Clutter is a matter of how much the content gets in the way of users instead of facilitating their path to the information they want.

Even after readers have purchased your product, they are still buyers when it comes to your information. They still have to buy into the idea that your information set is going to be easy to use and contain the information they need. If you show them a Frankenbook help system or some complex navigation system based on classifications they do not already have clearly in mind, you're not sending the right message. Your information set may be big and complex, but no matter where readers land in it, it needs to feel small and simple.

Adopting Every Page is Page One information design allows you to create a mammoth documentation set that feels small, safe, and comfortable.

Wikipedia has that small and simple feel on every page even though it is actually mammoth. And, despite being mammoth, it is easy to move around in. So is Amazon. So is YouTube. Why? Because they all use bottom-up navigation and because every page is a hub of its immediate area in subject space. Adopting Every Page is Page One information design will allow you to create a mammoth documentation set that feels small, safe, and comfortable.

EPPO and DITA

I have said some things in this book that are critical of certain aspects of DITA, particularly the idea, commonly attached to DITA, that dividing content into concept, task, and reference topics constitutes an adequate topic-based information design

(Chapter 9). You may be wondering if I am proposing EPPO as an alternative to DITA. The answer is that EPPO and DITA are orthogonal to each other. EPPO is an information design pattern. DITA is, in the words of one of its creators, Don Day, "basically just a generic but extensible markup standard."[11] You can implement EPPO using DITA or other tools. You can use DITA to create EPPO topics or to create manuals, hierarchical help systems, or Frankenbooks.

There are a number of advantages, both design advantages and process advantages, that come with topic-based authoring. There are process efficiencies to be gained by working in small units and avoiding the bottlenecks that come with trying to push large manuals through an approval and publishing system. There are opportunities for reuse. But both the advantages and the costs vary depending on which form of topic-based authoring you use.

DITA supports the use of building-block topics and other techniques that can greatly increase reuse, but at considerable costs in infrastructure and ongoing maintenance. Most DITA advocates seem to agree that you need to have a least five writers in an organization and significant opportunities for reuse before the savings outweigh the costs. On the other hand, you can do EPPO information design in a wiki for much less and reap many of the benefits of working in smaller units, though without the degree of reuse a DITA CMS might offer. You can also do EPPO in a DITA CMS using building-block topics, and other DITA features, to achieve greater levels of reuse.

While DITA is "basically just a generic but extensible markup standard" and, therefore, can be used to construct a wide variety of systems, any DITA CMS that you buy from a vendor will carry with it a particular set of built-in content design and delivery process biases. There is a huge difference between what someone with the right skills could build based on DITA's extensible markup standard, what the DITA Open Toolkit provides out of the box, and what a particular commercial DITA CMS implements. Make sure you consider what a particular system actually does, not what DITA makes theoretically possible.

---

[11] This quote about DITA comes from a comment made by Don Day on the Content Strategy Google Group. https://groups.google.com/d/msg/contentstrategy/XG5O1g4k1wk/nt8vsxfG208J

Most DITA advocates will agree that when you move to DITA, you need to put a lot of thought into how you design your information set and how you design your topics. Failing to do so, or relying on mechanically-separated topics generated from previous book-structured material, will usually lead to the production of Frankenbooks.

That does not mean that only one information design works with DITA. For example, with proper planning and design, you can use DITA to produce good manuals and hierarchical help systems. But if you're using DITA, I recommend you to choose an Every Page is Page One information design, first because it is the right design for modern technical communication, and second because it provides a needed design framework beyond the basic concept/task/reference topic-type trio.

EPPO and wikis

Wikis are a natural Every Page is Page One medium. Wikipedia, which is a poster child for the Every Page is Page One design pattern, illustrates how well the form fits the medium. Not only do wikis encourage Every Page is Page One design for individual topics, they are a natural bottom-up navigation medium. Wikipedia pages are peers. They are organized by linking and by grouping into categories (effectively, lists). Everything is a page, including the category groupings.

However, many companies do not install wikis because they are a great bottom-up Every Page is Page One medium. Instead, they install them because they are a great low-overhead, low-maintenance collaboration environment that the whole company can use to develop and share information. In many cases, documentation groups end up using a wiki not because they have chosen to adopt EPPO information design, but because they have been told to use the same tool that the rest of the company is using.

Since all tools have biases, and since the bias of a wiki is towards Every Page is Page One information design and bottom-up navigation, it can be challenging to produce traditional manuals and help content in a wiki environment. And while some wiki platforms do have some add-on support for this kind of output, it is not as natural or as complete as a tool designed for top-down information design and navigation.

If you are currently using a wiki (or, for that matter, a Web CMS) for documentation, switching to an Every Page is Page One model will immediately make your life easier because you will be using these tools with their bias rather than against it.

If you have decided to move to an EPPO information design (and I hope by this point in the book that you have!) then a wiki is an excellent tool choice if you do not want to go down the structured writing route.

Making the case for technical communication on the Web

Throughout this book, I have stressed that the EPPO design pattern is appropriate for most technical communication, whether it is on the Web or not, because EPPO supports how users of products have always looked for help and because EPPO is the model of information consumption that readers who live and work in the context of the Web apply to all information sources.

However, I have also pointed out the advantages of having technical information available on the public Web. This section provides suggestions on how you can make the case for putting your technical communications on the Web and how you can answer some of the objections you may face.

In terms of increasing the efficacy of technical communications, the case for the Web is clear. But technical communication groups often meet other objections when they try to move their content to the web. The potential objections from the marketing department have been noted above, and I have discussed how an Every Page is Page One approach can help alleviate those concerns. Here are some other objections and ways to handle them.

Competitors will steal our ideas

This one has always mystified me a little, but I have heard it from several sources. Some companies fear that if they put their documentation on the Web, their competitors will copy their products. Here are some ways to answer this objection.

- First, make sure the people you are talking with understand what you mean by *documentation*. If they think you mean engineering drawings or product specifications, then of course they don't want them on the Web. Explain that you are talking about end-user documentation, which, in most cases, is not supposed to contain trade secrets. There are some exceptions to this. For example, companies that sell components that are integrated into products sold by their customers may have to expose trade secrets to make those components usable. In that case, you need to take extraordinary measures to make sure your customers keep the documentation confidential.
- Second, ask what it would take for a competitor to get hold of your documentation today. In most cases, it would be trivially easy. Even for large and expensive systems, sales people for rival companies can get all kinds of information from their product champions within your customer's organization.
- Show them what results come up today when people do a Web search for help with your product. Those queries will often land on sites your executives would rather they did not land on, whether it be competitors, critics, or people giving bad advice. Explain that if your technical documentation was on the Web in a Web-friendly format, people would not hit those other sites as frequently.
- Make a list of leading companies that have their documentation on the web. Start with the likes of Google, Microsoft, and Apple, and add any companies in your space that your execs will be impressed with.

Our users prefer PDF

I'm never quite sure if this objection is serious or if it is just a way of brushing aside an issue that the executive isn't really interested in. If it comes from a sincere belief that users want PDFs, then the section titled "EPPO and PDF/help" (p. 237) provides some answers. But my suspicion is that this objection is not so much about PDFs in particular as about a belief that the status quo is good enough.

I have stressed throughout this book that the audience for technical information is changing. Traditional forms of documentation are losing readers to the Web. When executives claim that users are happy enough with the current documentation, the arguments in Part I, "Content in the Context of the Web," may help you make the case that things have changed and the status quo no longer holds.

Of course, there is still the possibility that your executives may not care that the company's documentation is losing readers, which brings us to the next objection.

No one reads the documentation anyway

If by "reads the documentation," they mean sits down and reads the manual like a novel, then this objection is entirely true. But that's not the point. The point is that people do read documentation – or, at least, technical information – when they get stuck and need help. We can establish this beyond doubt by surveying the forums on every topic under the sun where people ask for and provide help to each other on the Web. People seek out and read technical information when they have a specific problem and need help.

The question then is, does your organization care whether people get that help from your content or from someone else's. The knee-jerk answer to that question may be no, but ask some of the following specific questions and the answer may change:

- When customers search the Web with questions about our product, whose content do we want them to land on?
- When customers search the Web to resolve problems with our product, whose content do we want them to land on?
- When potential customers search the Web trying to solve the kind of problem our product solves, whose content do we want them to land on?
- When a competitor's customers search the Web trying to solve a problem in our field, whose content do we want them to land on?
- When technical journalists or bloggers search the Web looking for information about the features of products in our field, whose content do we want them to land on?

Phrase it like that and the executives you are talking to may start to connect your content to dollar signs.

You will, of course, need to make a strong and specific business case, which is outside the scope of this book. However, I will suggest that this is not the time to be meek. When you have a user population that lives and works in the context of the Web and expects easy access to the information they need the moment they need it, professional technical communication is more valuable than ever before and can generate more shareholder value than ever before. Or it can become irrelevant; it's your move.

The status quo in information design and information delivery is not a viable place to remain. We live in a world were every page is page one. It is no exaggeration to say that we must adapt or die.

CHAPTER 23

Afterword: EPPO, but Not for Everything

When you advocate for anything, there is always the danger that you will deceive yourself into thinking that the thing you care about applies universally. Of course, almost nothing applies universally, especially in the business world, and I have always been a believer in the right solution for the right business problem.

There is also a danger that when you talk a lot about something, people will think you are suggesting it should be used for everything and will propose cases where the thing you are advocating won't work, as if those cases invalidated the entire argument. But the truth is, there are few panaceas, and certainly none when it comes to the difficult business of human communication.

EPPO is not for everything.

EPPO is not for everything.

Over the course of this book, I have leveled criticism at books and book-like things, such as the typical book-derived help systems so common today. Books don't work well for a large percentage of technical communication tasks, especially for readers whose expectations for information delivery are conditioned by living and working in the context of the Web. But that does not mean there is no role for books in technical communication, and certainly not that there is no role for books in general. After all, this is a book you're reading.

There is a difference between documenting a product and documenting an idea. Most of technical communication is about documenting a product or a service: something tangible you can touch and poke. A lot of the other things we search the Web for are like this as well. But there are things it is harder get your hands on and play with, either because they are too distant or because they are intangible.

It is harder to play with a half-formed idea than with a half-understood product. The resistance to taking up new ideas is correspondingly higher, and the experimentation and error recovery method does not apply because there is no form of error reporting built into an idea. You can see this with something like minimalism: how do you know if you've implemented minimalism correctly? Unless you set up A/B testing similar to what John Carroll did, it is difficult to evaluate success or failure. For this reason, too, people often tend to fall back to a simpler version of the idea – which is why there has been a steady stream of articles and blog posts describing misconceptions about minimalism ever since *The Nurnberg Funnel*[8] was published.

Similarly, Every Page is Page One is an idea – specifically, a design pattern. You can certainly go out and write Every Page is Page One topics. In fact, there's a pretty good chance you have written some already. There are also plenty of EPPO topics out there you can read. But if you write an EPPO topic incorrectly, your authoring application is not going to beep and display an error message. One of the reasons I like to use rhetorically and computably structured writing for writing EPPO topics is that it allows you to build in some feedback. But still, you can only really test an information design pattern by doing A/B testing, and that is expensive. So it is not like you can drop this book after five minutes and come back the first time you see an error message.

Part of what I have tried to do in Part II, "Characteristics of Every Page is Page One Topics," is to provide a set of reasonably concrete benchmarks to evaluate your EPPO topics against. But all of these measurements are, to one extent or another, judgment calls. Your experience, and the feedback you get from your readers, will help hone your judgment in these matters, but you need a fair amount of theoretical grounding just to get to that point.

This is the sort of thing that people need a book for. It is why John Carroll wrote a book about how learners don't read books. It is why David Weinberger wrote two books about why the book is an inadequate vessel for knowledge. It is why I wrote this book about how to stop writing books and start writing Every Page is Page One topics. It is why you should still write a book too, if you are dealing with this kind of subject matter.

But technical communication, product documentation in particular, is not usually this kind of subject matter. The reader is a user and has something to use. The documentation necessarily takes a back seat to learning through use. And in the rich information foraging environment provided by the Web, people's natural desire for short, purpose-oriented information snacks is well fed, which affects their reading habits and expectations not only when they are on the Web, but with all the information sources they use – all of which, of course, they use in the context of the Web.

This does not mean that every discussion of ideas belongs in a book. Indeed, the development of ideas requires the cooperation and counterpoint of many minds. As David Weinberger argues, "Long form thought is not wide enough for deep thinking, big problems and big ideas require the counterpoint of many minds, not the monologue of one mind."[1] A web of topics from many authors may be required to develop and do full justice to a big idea.

This book would not have been possible if I hadn't aired these ideas on my blog over the last two years, or if I hadn't received corroboration, correction, and criticism from my readers on the blog. But at the same time, the ideas there developed piecemeal. At a certain point it became necessary to satisfy myself that it all hangs together, that it works as a coherent whole. Writing a book was a good way to figure that out. And if you have followed my blog over the last couple of years, you will definitely find things in the book that don't agree entirely with what I said in the blog, and you will certainly find that I have refined and changed a lot of the terminology I used in the blog. A blog is a great way to develop an idea. A book is a great way to consolidate and validate it.

In the development of thought, there is also the need for recapitulation. As fruitful as it may be, the web of many minds can become chaotic. It can fall into patterns of repetition. It can analyze endlessly and brilliantly, but it can have trouble with synthesis. There are points in the development of any argument when all concerned can benefit from a recapitulation that weighs all that has been said, then sifts, sorts, and synthesizes it. The recapitulation is not an end point. On the contrary, it is often a way to move the discussion forward by providing either a new point of departure or a new focus of debate.

In the development of thought, there is the need for recapitulation.

The act of recapitulation has value for the author as well. However steeped we may be in a web of ideas, it is not until we set them down that we discover how they fit together or how they fail to connect. Recapitulation clarifies and refines the author's

[1] http://vimeo.com/48199408

thoughts. It also provides a solid target to shoot at if the author fails to make the connections properly.

Recapitulation is also useful to the learner, to the person who comes to the subject or problem long after the debate has begun. We seldom start our initiation into a subject with a program of sustained reading (though there are exceptions). Instead, after we have played around the edges for a while, we get to the point where we want to gain a broader understanding of what has been thought and said in the field, and for this purpose a recapitulation is in order. Thus after a period of tinkering and exploration, we reach for a book to launch us into the study of a field.

So, I am not saying the writing of books must be abandoned. But now that we live and work in the context of the Web, the role of Every Page is Page One topics has expanded greatly, and if professional technical communicators want to remain relevant in this world, they need to create more Every Page is Page One topics and fewer books and manuals.

Glossary

blocks

In the context of *Information Mapping*, an information block is a unit of content of a defined type that may be combined with other information blocks to create documents. Information blocks are similar to, though not necessarily identical to DITA topics.

building-block topics

In the context of this book, the term *building-block topic* is used to denote a unit of content that is designed to be combined with other pieces of content to create information products, but is not necessarily intended or designed to function as an independent information source for a reader.

bursting

Before being stored in a content management system (CMS), content is often broken apart into its individual components. This process of breaking content into its component parts is called segmentation or bursting (from *Managing Enterprise Content: A Unified Content Strategy*[24]).

computably structured

In this book, the term *computably structured* refers to a content data format that supports the processing of that content by algorithms. Technically all content created on a computer is computably structured, but the term is here reserved for formats that are designed to support multiple uses of the content and to be independent of any one application.

concept

In the context of *DITA*, concept is one of the three basic topic types, along with task and reference. In the context of *Information Mapping*, concept is one of the six types of information blocks. In the context of this book, a concept is a major foundational idea of a system or process that the user needs to grasp to use that system or process effectively.

content API

A *content API* is an interface that is used to deliver content to computer programs, usually over a network and particularly over the Web. There is a sense in which every URL is a content API, but the term is generally reserved for systems where the URL contains parameters that define the content to be retrieved, rather than simply specifying a static location. Twitter is an example of a content API which delivers a stream of tweets from the people you follow, based on your identity when you log in.

database

Any collection of data that can be queried in a reliable fashion. The term is often used to mean only content stored in a formal database management system, and sometimes specifically a relational database management system, but in fact any query-able data-store is a database, including, for example, an XML document.

derived purpose

In the context of this book, the term *derived purpose* is used to describe a *purpose* that a user determines is necessary as one part of achieving an objective. For example, it has been said that users are not interested in drills, they are interested in holes. But a person who wants to create a hole may, as part of that objective, want to attach a drill bit to a drill. Attaching the drill bit is a derived purpose. People often search for help with their derived purpose rather than their ultimate purpose.

DITA

DITA (Darwin Information Typing Architecture) is an XML architecture for designing, writing, managing, and publishing information (see dita.xml.org).

dynamic semantic clustering

In the context of this book, a semantic cluster is created when a Web search or an act of curation brings together pieces of content from diverse locations. Dynamic semantic clustering describes what happens when a user performs a search or accesses a content API. In that case, pieces of content are drawn together dynamically based on the search terms the user chose. More and more, people experience the Web not as a static set of pages or sites, but as a sequence of dynamic semantic clusters.

Every Page is Page One

Every Page is Page One is an information design pattern that starts with the recognition that readers move frequently from one information source to another. It seeks to accomodate this behavior by creating content that works as page one for every reader no matter how that reader arrived at that content.

Frankenbooks

In the context of this book, I use the term *Frankenbook* to refer to an online help system or information set, usually composed of multiple source books, or compiled from a large number of *building-block topics* with a huge table of contents that is difficult or impossible to navigate due to its size and lack of overall organization, particularly one in which the individual topics do not work well when viewed on their own.

generic

In the context of this book, a *generic* topic is one that has no distinctive or subject-specific rhetorical structure.

information foraging

Information foraging is a term used to describe how readers look for information, likening it to the way wild animals forage for food. See [9] and [21].

Information Mapping

Information Mapping® is a proprietary information design pattern based on the idea that good content can be created by assembling content blocks of certain defined types into maps that define documents (see *Information Mapping*[15]).

long tail

A *long tail* is a statistical distribution in which a large number of items occur far from the mean. In the context of information, it describes a situation in which there are many pieces of information which individually are of interest only to a small number of people, but which together account for as much of the total information requirements as the few pieces in high demand.

motive

In the context of this book, the word *motive* is used to describe the readers original reason for wanting to do a task.

page

In the context of this book, a *page* is a unit of display for information – a page in a book or a Web page. By contrast, a unit of information of interest to a reader is called a *topic*. A topic may or may not occupy a single page.

pathfinder

In the context of this book, a *pathfinder* topic is one that is designed to help readers find their way through a process or technology or relate the process or technology to their business problem.

purpose

In the context of this book, the purpose is what a user is trying to achieve when looking for information. Essentially, when you ask users what they are trying to achieve, their answer is usually their *purpose*. The *purpose* will usually be different from the *motive*, since people will normally answer this question in terms of the immediate practical problem they are trying to solve, rather than their original reason for acting.

rhetorically structured

In the context of this book, the term *rhetorically structured* describes content that follows a deliberate plan or template designed to best express a particular idea or explain a particular subject. While all content has some form of rhetorical structure, the term is used here to describe a structure that is defined for a whole class of content and is used to guide the writing of instances of that class.

RTFM

A phrase commonly used in technical communications and support. The polite expansion of the acronym is "Read The Fine Manual." Other expansions are also common.

satisficing

Satisficing is a decision-making strategy in which a person chooses an option that is good enough given the energy expended rather than an optimal solution that would require more energy. Most people's information seeking strategy is satisficing: they will stop looking when they find information that is good enough rather than continuing to spend time and money looking for perfect information.

soft linking

Soft linking is a technique in which links are generated during the publication process from *subject affinities* noted in the source text. Subject-affinity notation records the relationship of a word or phrase to a real-world object, rather than a link to a specific information resource. Because the source text contains no explicit links, there are no links to manage and no links to break.

structured writing

Structured writing is a writing strategy that begins by defining the content, order, and form of a topic prior to writing the topic itself. It includes both *rhetorically structured* and *computably structured* writing.

subject affinity

A subject affinity is a relationship between the subject of a particular topic and other subjects that are mentioned in the topic. A subject affinity is not a link or a related topic. It is an affinity between subjects themselves, not between content that describes subjects. *Computably structured writing* can be used to capture the subject affinities of a topic, which can then be used for management and linking.

topic

Two distinct meanings of the word topic are used in technical communication today. One meaning refers to a short, independent unit of information that does not rely on other information products. Examples include blog posts and magazine articles. I call these *Every Page is Page One* topics. The other meaning refers to a discrete block of content that is intended to be used to build a larger piece of content. For example, in a system like *DITA*, you might create a topic that is used to build one or more books. In this book, I call these *building-block* topics. If I use the term *topic* alone, I mean Every Page is Page One topics, unless the context clearly shows otherwise. Note that there is no debate here about what a *topic* really is. I'm just using the same word for two related but distinct concepts. You could use building-block topics to build Every Page is Page One topics. You don't have to pick sides, you just have to understand the distinction.

up hill

In the context of this book, the phrase *up hill* is used to describe a piece of computably structured content that captures the semantics of its subject matter in a way that permits the content to be transformed into other structures with more presentational semantics (but usually less subject semantics). It is easy to transform from subject semantics to presentational semantics, but not the other way round. Thus content with more subject semantics is *up hill* from content with fewer subject semantics, or with none. The more up hill the content is, the more opportunities there are to use, transform, and deliver it in useful ways.

Bibliography

[1] Anderson, Chris. *The Long Tail: Why the Future of Business is Selling Less of More.* Hyperion.

[2] Anderson, Chris. "Consumer Surplus in the Digital Economy: Estimating the Value of Increased Product Variety at Online Booksellers." *Wired* 12, no. 10 (October 2004). http://www.wired.com/wired/archive/12.10/tail.html.

[3] Arel, Ena. "Tips and Tricks: Getting from Obvious to Valuable Technical Content." http://techwhirl.com/obvious-to-valuable-technical-content/.

[4] Brooke, Andrew. "Topical Docs." *A Tech Writer's World: The science and philosophy of technical communication.* http://techwriters-world.blogspot.com/2011/06/-topical-docs.html.

[5] Brynjolfsson, Erik, Yu (Jeffrey) Hu, and Michael D. Smith. "Consumer Surplus in the Digital Economy: Estimating the Value of Increased Product Variety at Online Booksellers." *Management Science* 49, no. 11 (November 2003). June 2003 version available at SSRN: http://papers.ssrn.com/sol3/papers.cfm?abstract_id=400940.

[6] Brynjolfsson, Erik, Yu (Jeffrey) Hu, and Michael D. Smith. "The Longer Tail: The Changing Shape of Amazon's Sales Distribution Curve." September 20, 2010. Available at SSRN: http://papers.ssrn.com/sol3/papers.cfm?abstract_id=1679991.

[7] Carr, Nicholas. *The Shallows: What the Internet is Doing to Our Brains.* W.W. Norton. 2011.

[8] Carroll, John. *The Nurnberg Funnel: Designing Minimalist Instruction for Practical Computer Skill.* Cambridge, MA: MIT Press, 2007.

[9] Chi, Ed H., Peter Pirolli, Kim Chen, and James Pitkow. "Using Information Scent to Model User Information Needs and Actions on the Web." http://www2.parc.com/istl/-groups/uir/publications/items/UIR-2001-07-Chi-CHI2001-InfoScentModel.pdf.

[10] Farkas, David K. "Layering as a Safety Net for Minimalist Documentation." In *Minimalism Beyond the Nurnberg Funnel,* edited by John M. Carroll, 247–270. Cambridge, MA: MIT Press. 2003.

[11] Geiger, Brian J. "Tarragon Mac and Cheese Recipe." 2008. http://thefoodgeek.com/blog/-2008/12/11/tarragon-mac-and-cheese.html . © 2008, The Food Geek, CC by 3.0.

[12] Gentle, Anne. *Conversation and Community: The Social Web for Documentation.* 2nd ed. Laguna Hills, CA: XML Press. 2012.

[13] Gladwell, Malcolm. *The Tipping Point.* Little-Brown. 2000.

[14] Heath, Chip and Dan Heath. *Switch: How to change things when change is hard.* Random House Canada. 2010.

[15] Information Mapping International. http://www.informationmapping.com.

[16] Levine, Rick, Christopher Locke, Doc Searls, and David Weinberger. *The Cluetrain Manifesto: The End of Business as Usual.* Basic Books. 2001. Available online at: http://cluetrain.com/book/hyperorg.html.

[17] MacInnis, Peter. *The Speed of Nearly Everything: From Tobogganing Penguins to Spinning Neutron Stars.* Pier 9. 2008.

[18] Mamykina, Lena, et al. "Design Lessons from the Fastest Q&A Site in the West." Author's version: http://bid.berkeley.edu/files/papers/mamykina-stackoverflow-chi2011.pdf.

[19] Mayer-Schönberger, Viktor and Kenneth Cukier. *Big Data: A Revolution That Will Transform How We Live, Work, and Think.* Houghton Mifflin Harcourt. 2013.

[20] Nesbitt, Scott. "It's help, but not (quite) as we know it." *Communications from DMN.* http://www.dmncommunications.com/weblog/?p=2605.

[21] Nielsen, Jakob. "Information Foraging: Why Google Makes People Leave Your Site Faster." Nielsen Norman Group. 2003. http://www.nngroup.com/articles/information-scent.

[22] OASIS. *OASIS Darwin Information Typing Architecture (DITA) Technical Committee.* https://www.oasis-open.org/committees/tc_home.php?wg_abbrev=dita.

[23] Parnin, Chris, et al. "Crowd Documentation: Exploring the Coverage and the Dynamics of API Discussions on Stack Overflow." *Georgia Tech Technical Report* GIT-CS-12-05. http://larsgrammel.de/publications/parnin_2012_crowd_documentation.pdf.

[24] Rockley, Ann and Charles Cooper. *Managing Enterprise Content: A Unified Content Strategy.* 2nd ed. New Riders Press. 2012.

[25] Statistic Brain. "Google Annual Search Statistics." http://www.statisticbrain.com/-google-searches/.

[26] Wachter-Boettcher, Sara. *Content Everywhere: Strategy and Structure for Future-Ready Content*. Rosenfeld Media. 2012.

[27] Weinberger, David. *Too Big to Know: Rethinking Knowledge Now That the Facts Aren't the Facts, Experts Are Everywhere, and the Smartest Person in the Room Is the Room*. Basic Books. 2011.

[28] Weinberger, David. *Everything Is Miscellaneous: The Power of the New Digital Disorder*. Holt Paperbacks. 2007.

[29] Womack, James P. and Daniel T. Jones. *Lean Thinking*. 2nd ed. Simon & Schuster. 2003.

Index

A

affinity (*see* subject affinity)
aggregation, 20
agile methodologies, 233–235
  test-driven development and, 214
Amazon
  bottom-up navigation, 51, 245
  faceted navigation on, 44
  as a filter, 22
  lists on, 67, 111, 219
  long tail, 14, 242
Anderson, Chris, 11
API, 45
  computable structures, 194
  content, 204
  crowdsourced documentation, 22
  example, 98–100, 180
  faceted navigation, 112
  function reference, 103
  reuse, 225
  soft linking, 217
Apple, 243
application programming interface (*see* API)
Arel, Ena, 93
Atherton, Mike, 28
Atlassian, 121
  attribution on topics at, 161
authority versus experience, 18
AutoCatch, 44, 111

B

backwards navigation, 179–180
Bezos, Jeff, 242
big picture topics, 167–175
books
  assembled from topics, 75, 240
  need for, 254
  organization, 25, 31
bottom-up navigation, 60–67, 245
  findability and, 142
  qualifications and, 136
  structured writing and, 205
  subject affinities and, 241
bottom-up organization, 51
  linking and, 159
  SPFE and, 194
  structured writing and, 205
Brooke, Andrew, 71
building-block topics, 71–73
  context-dependent, 72
  context-free, 73
bursting
  Frankenbooks, 42
  help system, 75
business case, 227

C

Carroll, John, 4, 87
  guided exploration cards, 182
  manual preference studies, 240

on minimalism, 181, 186, 254
on procedures, 106
on qualifications, 129
on sense making, 235
on tutorials, 162
styles of learning, 148
change, managing content, 230
checklists, 109, 178
chunking
chunk size, 210
help system, 75
mechanical, 73
citations versus subject affinities, 59
classification, 41, 47
card sorting, 49
faceted, 43–45
limits, 45, 65
schemes, 48
subject affinity and, 58
top-down navigation and, 47
vs. curriculum, 34
clustering, dynamic semantic, 26–29, 65
CMS (*see* content management systems)
commercial purpose, 104
component content management systems (CCMS), 193, 228
computably structured writing, 192–195
advantages, 200
alternate encodings, 194
reuse and, 225
concept topics, 105–108, 113–115
conditional text, 225
configuration tasks, 109
Confluence, 121
conformance, quality and, 201
content
aging, 231
data-structured, 199
distributed nature, 25–29
duplicate, 222
exchanging, 204
future proofing, 202–203
high demand, 14–15
interoperability, 204
lifecycle, 231
management, 235–237
managing change, 230
manipulating structured, 202
obsolete, 231
off-Web, 60
content API, 204, 223
interactive, 226
reuse, 225
content management systems, 193, 207, 236
DITA, 247
content marketing, 241–246
content shifting, 20
context
building-block topics and, 72
content shifting and, 20
disambiguation, 121
documenting, 94–95
EPPO design and, 74
establishing, 55–58, 78, 117–122, 136, 155
establishing with a graphic, 119
establishing with a link, 121
example of setting, 118
links and, 143
reuse and, 73
search and, 2, 121
specific purpose and, 86
statement, 130, 156
Web, 71
continuous delivery, 229
cookbooks, levels in, 132
Cukier, Kenneth, 63
curation, 20
curriculum versus classification, 34

D

databases
  computable structures and, 193
  reference information and, 112
Day, Don, 247
decision support, 92
democratization of knowledge, 141
dependencies, reader versus subject, 125
derived purpose, 88–90, 125
design patterns, 198
  computably structured writing and, 200
detection theory, 98
disambiguation, 121
DITA, 41, 75, 204
  cost of implementation, 247
  EPPO and, 246
  maps, 107
  Open Toolkit, 247
  reuse, 203, 224
  structural type, 196
  topic types, 106
DocBook
  conversion to, 204
  format independence and, 203–204
  structural type, 196
  structured content example, 200
Drupal, 193
duplicate content, 222
dynamic semantic clustering, 26–29, 65

E

ebooks, finding information in, 11
Eckel, Bruce, 133
embedded user assistance, 148
EPPO design
  context, 74
  limitations, 253
  linking, 140
  minimalism, 184
  purpose and, 151
  self-contained, 79
  specific purpose, 85
EPPO topics (*see* topics)
EPPO-simple, 196–197
experience versus authority, 18

F

Facebook, as a filter, 22
faceted navigation, 43–45, 48
  for APIs, 45, 112
  lists and, 67
  on Amazon, 44
  on AutoCatch, 44
FAQ, origins of, 76
Farkas, David K., 186
file formats, 192–194
  structured content and, 195
filtering, **9–23**, 28
  examples of, 21
findability
  content aging and, 231
  information scent and, 83
  irregular affinities and, 58
  linking and, 142
  purpose and, 95
  qualifications and, 129–130
flattening, 60–64
Flickr, 42
foraging (*see* information foraging)
format independence, 203
formats, 195
  (*see also* specific formats by name)
  microformats, 204
  open vs. closed, 202
  structured, 195
FrameMaker, 153, 196, 204, 240
Frankenbooks, 41–42
future proofing, 202–203

G

generic topics, 113–115, 154
Gentle, Anne, 19
Getting Started guides, 172–173
Gladwell, Malcolm, 17
glossary of terms, 257–262
Google
  App Engine example, 170–172
  as a filter, 21
  Chrome documentation, 73
  Datastore documentation example, 118
  duplicate content and, 222
  information foraging, 1
  less popular results, 122
  usage, 10
guided exploration cards, 182
guides
  Getting Started, 172–173
  Popular Mechanics example, 34
  style, 153

H

Hackos, JoAnn, 237
Hall, Brian S, 239
HAT (help authoring tool), 73
Heath, Chip and Dan, 5
help
  assembled from topics, 241
  EPPO and, 237
  tri-pane, 32
help authoring tool (HAT), 73
hierarchical organization, 37, 73
  cultural bias towards, 39, 149
  limitations, 38
  links subvert, 142
Hopkins, Rebecca, 179
HTML5, 200
hubs
  lists as, 67
  navigation, 64
  topics as, 28, 60

I

indirection, 217
information architecture
  bottom-up, 51–67, 140
  EPPO, 140
    (*see also* EPPO design)
  top-down, 31–49
information foraging, 1–3, 140
Information Mapping, 75
  information block types, 106
  maps, 107
  structured writing and, 190–192, 196
information scent, 1–3, 48, 83
  EPPO topics and, 71
  topic type and, 100
interactive pages, and reuse, 226
interoperability, 204
isomorphic drawing, 62

J

JavaDoc, 194
Jobs, Steve, 161, 238
Johnson, Tom, 63
  on chunking, 210
  on help models, 33
  on shape of help, 106
  on topic size, 85
  on topic-based authoring, 93

K

Kay, Michael, 15

L

learning styles, 148

levels, 131
  books and, 132
  changing, 131
  hierarchical, 38
  qualifications, 127
  staying on one, 78, 157
lifecycle, content, 231
limitations of EPPO, 253
LinkedIn, as a filter, 22
linking, 215–219
  as alternative to reuse, 226
  bottom-up navigation and, 51, 142
  context setting and, 143
  crowdsourcing and, 215
  disambiguation, 121
  EPPO design, 140, 159
  indirect, 217
  soft, 215, 232
  structure and, 201
  subject affinity, 54–59
  videos and, 164–165
lists, 66–67
  on Amazon, 67
  as hubs, 67
  role of, 66
  soft linking and, 219
  vs. tables of contents, 66
  on Wikipedia, 67
long tail, 11–17
  Amazon, 14, 242

M

MacInnis, Peter, 39
Manicouagan crater example, 54–58
maps
  DITA, 107
  Information Mapping, 107
March, James, 5
MathML, 197
Mayer-Schönberger, Viktor, 63
McGovern, Gerry, 95, 241
Mercator projection, 61
metadata, 119–120, 207–214
  as navigation, 207
  content aging and, 232–233
  EPPO topic, 212
  establishing context with , 155–156
  meaning of, 208
  structured writing and, 214
  types of, 209
  use of the term, 209
  when to write, 151
  XML, 208
methodologies
  agile, 233–235
  structured, 104, 189
Meyers, Peter J., 222
microformats, 204
Microsoft Word, 153, 195, 240
minimalism, 106, 181–187
  big picture and, 168
  EPPO design, 184
  versus comprehensive documentation, 185
motive, 86–89

N

navigation
  backwards, 179–180
  book, 31
  bottom-up (*see* bottom-up navigation)
  faceted, 43–45
  large information sets, 64
  lists in, 66
  multidimensional, 63
  page-based, 51
  top-down (*see* top-down navigation)
Nesbitt, Scott, 73
Nielsen, Jakob

guidelines for writing content, 222
information foraging, 1–3
NPR, reuse at, 223

O

organization
book, 31
bottom-up (*see* bottom-up organization)
hierarchical, 37
limitations, 38
physical world constraints, 40
representing multidimensional, 63
top-down, 31
two-dimensional, 60
overviews, 168
writing, 170

P

Parnin, Chris, 22
pathfinder topics, 115, 173–175
PDF, 237–241
PDF, user preference for, 250
personality, brand and, 161
perspective drawing, 62
Pinterest, as a filter, 22
Pirolli, Peter, 1
Popular Mechanics, 34
prerequisites
reader, 141–142
Pringle, Alan, 237
procedures, 90
form of, 97
Information Mapping, 106
minimalism and, 184
tasks versus, 93, 108–110
workflows and, 178
process
continuous delivery, 229
productivity
unit size and, 228
programming tasks, 109
projections, 61
purpose, 86
commercial, 104
derived, 88–90
documenting a topic's, 151
findability and, 95
related to features, 88
specific and limited, 135, 151
user vs. topic, 90

Q

qualifications, 136
dependencies and, 125
findability and, 129–130
contrast with judgments, 129
topic types and, 157
qualified readers, 78, 123, 127
becoming, 141–142
writing for, 156–157
quality
conformance and, 201
subject-specific markup and, 204
Quora, 18

R

reader behavior, non-linear, 181
reader dependencies, 125
readers
qualified, 123, 127
unqualified, 141–142
RecipeML, 197
recipes
as topic type, 97
Tarragon Mac and Cheese, 80
Reddit, as a filter, 22
reference topics, 105–108, 110–113
references, 161

as link targets, 162
reuse, 203, 221–226
context and, 73
interactive pages and, 226
on the web, 221
reuse strategy
paper-style vs. Web-style, 225
rhetorically structured writing, 190–192
ROI, 227
RSS
as a filter, 22

S

satisficing, 16
search
impact of failure, 15
imprecise, 121
limitations, 17
micro- vs. macro-level, 185
optimization, 222
satisficing, 16
semantics
preserving, 202
SEO (Search Engine Optimization), 222
sequences of topics, 177
SGML, 202
Shapes of Help figure, 107
single sourcing, 203
Smartblogs.com, 160
soft linking, 215–219, 232
example, 218
indirection and, 217
SPFE, 194
Stack Overflow, 11, 18, 22, 28
Stanchak, Jesse, 160
structure
book, 31
formal, 196
general, 195
hierarchical, 149
(*see also* hierarchical organization)
subject-specific, 196
types of, 195
structured writing, 189–205
bottom-up organization and, 205
computable, 192–195
concerns with, 189
importance of, 190
metadata and, 214
rhetorical, 190–192
SPFE, 194
varieties of, 190
style
uniformity of, 159
style guides, 153
subject affinities, 53–59
bottom-up navigation and, 241
citations versus, 59
irregular, 58
linking and, 215
subject dependencies, 125

T

tables of contents (*see* TOCs)
Tarragon Mac and Cheese recipe, 80
task switching, dangers of, 151
tasks
configuration, 109
DITA, 108
procedures versus, 93, 108–110
programming, 109
sequence of, 177
topic type, 76, 105–108
workflows and, 178
writing, 86–88, 94
taxonomies
locally managed, 58
technical communication on the Web, 249

TechWhirl, 93
textbook model, 227
    user assistance versus, 147–150
TOCs
    as metadata, 208
    expressing sequences with, 177
    function of, 34
    limitations, 32
    role of, 168
Tognazzini, Bruce, 243
tools, 87
    closed, 195
    for structured writing, 192
    moving to new, 87
    selecting, 193
top-down navigation
    limitations, 65
    strengths, 47
    when to use, 65
    with bottom-up, 66
topic types
    artificial vs. natural, 97
    concept, task, and reference, 105–108
    conforming to, 152
    creating formal, 102–105
    defining, 103, 152–155
    defining qualifications in, 157
    discovering, 103
    EPPO, 97–115
    evolution of, 100
    Information Mapping, 106
    XML, 104
topics
    as hubs, 60
    big picture, 167–175
    building-block, 71–73
    commercial purpose of, 104
    concept, 113–115
    concept versus generic, 113
    defining the purpose of, 89
    definition, **71**
    derived purpose, 125
    economics and the evolution of, 75
    Every Page is Page One, **73–75**
    generic, 113–115
    levels, 131
    pathfinder, 115, 173–175
    presentational, 73
    reference, 110–113
    scope, 85
    self-contained, 79
    sequence of, 177
    size of, 92, 184
    structure, 102
    videos as, 165–166
    Web, 76–77
    workflow, 177
    writing, 150
    writing big-picture, 170
    writing EPPO, 147–166
    writing on one level, 135–137
    writing self-contained, 150
tri-pane help, 32
TurboTax, 197
tutorials, 162–163
Twitter, as a filter, 21
type (*see* topic types)

U

understanding, levels of, 127
Usenet, 76
user assistance
    embedded, 148
    textbook model versus, 147–150

V

version control, 193
videos, 163–166
    used as topics, 165–166

W

Wachter-Boettcher, Sara, 20, 28, 223
Wales, Jimmy, 242
Web
  how we use, 25
  information foraging, 1–3
  topics, 76–77
Web content management systems (WCMS), 193, 207
WebMD, 36
websites
  as filters, 65
  content aging, 231
  dynamically generated, 219
  tree structure, 25–29
Weinberger, David, 5, 18, 21, 142, 254
  on filtering, 9, 23, 28
  on linking, 139
  on metadata, 209
  on style, 161
Wikipedia, 28
  bottom-up navigation, 51, 245
  disambiguation, 121
  footers, 57
  lists on, 67
  sidebars in, 56
wikis, 248
Word, Microsoft (*see* Microsoft Word)
WordPress, 18, 193
  Codex, 89, 173–175, 180
  data format, 196
workflow topic, 177
writing
  structured, 189–205
    (*see also* structured writing)
  task-based, 86–88, 94

X

XML, 190
  (*see also* DITA)
  (*see also* DocBook)
  file as database, 112
  metadata, 208
  topic types and, 104

Y

YouTube
  as a filter, 22
  bottom-up navigation, 246
  linking, 164

Colophon

About the Author

Mark Baker is a twenty-five-year veteran of the technical communication industry, with particular experience developing task-oriented, topic-based content, and in designing and implementing structured authoring systems. He is also a frequent speaker on matters related to technical communications and structured authoring, and contributes to several publications in the field. Mark is currently President and Principal Consultant for Analecta Communications, Inc.[1] in Ottawa, Canada.

Mark's blog, Every Page is Page One[2] is focused on the idea that, in the context of the Web, Every Page is Page One, that the future of technical communication lies on the Web, and that to be successful on the Web, technical communicators cannot simply publish traditional books or help systems, they must create content that is native to the Web.

About XML Press

XML Press (http://xmlpress.net) was founded in 2008 to publish content that helps technical communicators be more effective. Our publications support managers, social media practitioners, technical communicators, and content strategists and the engineers who support their efforts.

Our publications are available through most retailers, and discounted pricing is available for volume purchases for business, educational, or promotional use. For more information, send email to orders@xmlpress.net or call us at (970) 231-3624.

[1] http://analecta.com/
[2] http://everypageispageone.com

Lightning Source UK Ltd.
Milton Keynes UK
UKOW02f2109251113

221836UK00007B/193/P